EARTH SCIENCE SERIES

No. 1

Edited by J. Sutton, D.Sc., Ph.D., D.I.C., F.R.S.
Professor of Geology, Imperial College, University of London
and J. V. Watson, Ph.D., A.R.C.S.
Imperial College, University of London

PHYSICAL PROCESSES OF SEDIMENTATION
An Introduction

JOHN R. L. ALLEN, D.Sc., B.Sc., F.G.S.
Reader in Geology in the University of Reading

Physical Processes of Sedimentation

AN INTRODUCTION

London
GEORGE ALLEN AND UNWIN LTD

FIRST PUBLISHED IN 1970

SBN 04 551013 x *cloth*
SBN 04 551014 8 *paper*

PRINTED IN GREAT BRITAIN
in 10 *on* 12*pt Times Roman*
BY WILLMER BROTHERS LTD
BIRKENHEAD

TO JEAN

PREFACE AND ACKNOWLEDGEMENTS

The study of ancient and modern sedimentary deposits for their own sakes has a long history, if a not wholly respectable one.

The originator of this attitude to sedimentary deposits is undoubtedly Henry Clifton Sorby (1826–1908), a Yorkshireman of independent mind, and also means, who devoted a long life to outstandingly original scientific research at his home near Sheffield, at which University there now exists a Chair of Geology bearing his name. We may justly call Sorby the Father of Sedimentology. He was firmly aware of the essential realities, that the problems of geological science exist in the form of deposits encountered in the field, and can be identified only by first making field observations. His descriptive work is of the highest calibre, but it was all the time governed by his knowledge of, and search for, processes. Sorby the geologist was essentially an environmentalist, who sought first to know, and then to understand in terms of mechanisms; an exploitation of his newly gained understanding almost invariably followed. For example, his remarkable experimental work on sedimentary structures due to current action, performed in the 1850's, led him to use the same structures as he found them in rocks to make novel environmental and palaeogeographical interpretations.

Sorby has had many direct if not always acknowledged scientific descendants, of whom perhaps P. G. H. Boswell, G. K. Gilbert, and W. H. Twenhofel have been the most influential, but since their day there have arisen other, and perhaps less laudable, attitudes than his towards the study of sediments. For it has been supposed by many that sedimentary deposits are merely to be described, or, to use the term fashionable amongst advocates of this view, characterized. These descriptions or characterizations, unilluminated by prior questions, have often been limited to just one or a few properties of the deposits examined. When interpretations are attempted, one finds as often as not that little reference is made to what we can see happening in the modern world, or can infer from the basic sciences on which much of geological science depends. It is no denial of the importance and place of observational work to say that such attitudes as these to geological problems are likely to prove bankrupt from the beginning. It is to my mind more in keeping with the scientific spirit to believe that the problems posed by sedimentary deposits can be solved only after blending broadly based observations with our knowledge of the basic sciences and the events of the modern world.

If sedimentary deposits have often been falsely approached, the environmental attitude itself has not always been clearly interpreted. To say, as the result of a comparison, that a particular set of beds is of fluviatile origin is merely to give a name to the environment of deposition. This naming of environments is an essential first step in the interpretation of any deposit, but it is neither a sufficient step nor, in the majority of cases, a desperately taxing one. The picture must be painted in greater detail and given flesh and blood: what kind of river was it, and how big and powerful, and where did it originate and flow to? These and other similar questions must be answered, quantitatively when possible, if the environmental hypothesis about the set of beds is to have any substantial predictive power. They are difficult questions, however, which oblige us at once to consider rivers as they now exist and the physical mechanisms that govern river flow.

It is considerations such as these that I have sketched in this book, in which I examine the physical background to the formation of terrigenous clastic deposits, and the way in which the properties of these deposits depend on the character and magnitude of the relevant physical processes. The result of my attempt is a synthesis, reaching over several disciplines, which I believe will appeal to undergraduates in their second or third year of study of geology and to post-graduates who are beginning to research in the sedimentological field. My synthesis assumes in the reader an introductory descriptive knowledge of sedimentary deposits, such as can be obtained from several recent books of high excellence in their field. I have not shirked the task of summarizing what I regard as the geologically significant features of the deposits I am seeking to explain physically, but I have naturally given most attention to developing the physical background required for their understanding, since it is the physics of sedimentation that is treated least satisfactorily, if touched on at all, in degree courses in geology. I believe that a grasp of the physical background to sedimentation should be cultivated by all whose work brings them into contact with sedimentary deposits. Our knowledge of sedimentary deposits will then be critical and predictive rather than encyclopaedic.

I know from my own years as a student, and later as an investigator in my own right, that the abstract and precise modes of thought that are second nature to the physicist and mathematician, are often foreign to the geologist faced with a highly complex nature and having behind him a largely descriptive training. But if we are to grasp the physical background to sedimentation, we also will have to cultivate the habits of precise and abstract reasoning, whilst continuing to recognize the uncertainties that nature poses. The goal is within reach, however, if a start is made at the beginning and common experience is at all stages appealed to in order to develop physical intuition. My Chapter 1 is an outline of the basic physics of fluids and sediment movement; it begins at the beginning, with a sketch of the nature of the problem and a statement of Newton's laws of mechanics on which all else depends. Many equations will be found in this as in later chapters, but there is nothing frightening or mysterious about these equations, and nothing beyond the mathematical equipment of a student entering a science department in a university. Each equation is a logical statement in symbols about a real physical situation, not about fairyland, and should be studied as carefully as any statement made in words. Once the terms of the equation are understood, the reader is the master of it and the possessor of its message and predictive powers. If difficulties are still experienced, graph the equation to bring it into subjection; but never throw in the sponge. Some of the difficulties arise from the unfamiliarity of the language. Appendix III should help comprehension.

The remaining chapters of the book are about the physical processes that are important in different major environments of deposition of terrigenous clastic sediments. All the chief regimes of wind and water are discussed, and my final chapter is about the realm of ice. I have tried to make my approach a uniform one, such divergences as will be found having been forced on me by the nature of a particular problem or the limitations of knowledge. Each chapter is concluded with a number of items for reading, some of which are standard works intended for reference, and others research papers of particular merit or importance in the field. The items for reading are not, however, intended to represent every topic discussed.

At a time when many countries are turning to the International System (Système Internationale d'Unités, abbreviated SI) of physical units for use in science and commerce, I have had to consider very carefully the question of whether to use this system in this volume, or to adhere to the old metric (CGS) units. The SI units are rational, coherent and comprehensive, as well as being logically superior to any other system. Research workers in the field of physical science will undoubtedly turn to their use with little delay. I believe the position to be quite different for geological readers who have yet to complete their degree courses and who, generally speaking, are not in close touch with physics and mathematics. In this book, therefore, I have decided to retain CGS units, on the grounds, firstly, that most of the texts and research papers I recommend for study employ these units (or some readily converted familiar set of units) and, secondly, that during the effective life-span of this volume the reader is likely to be more familiar with the CGS than the SI units. In this way, I believe, the difficulties associated with the transition to the new, internationally accepted system of units will be minimized. It is worth noting here that in the International System the metre is the unit of length, the kilogramme the unit of mass, and the second the unit of time. The unit of force is the newton (kg m s^{-2}), of energy the joule (kg m^2 s^{-2}), and of power the watt (kg m^2 s^{-3}). Reference to Appendix I, listing the CGS units used in this book, will allow conversion factors to be calculated whenever necessary.

I would like to record my great indebtedness to Dr B. D. Dore, Professor J. N. Hunt, and Dr A. D. Stewart, who have read chapters from this book and made many useful suggestions which have led to improvements; a further debt of thanks is owed to Professor Hunt for his permission to include in Chapter 4 a hitherto unpublished result on turbulent suspension. The facts and opinions expressed remain, however, my own responsibility. I would also like to thank those who have kindly permitted me to reproduce copyright materials; fuller acknowledgements of sources are made at appropriate places in the text. Also, my thanks go to Mrs D. M. Powell and Mrs B. Tracey for their careful preparation of the manuscript.

Finally, I thank Professor J. Sutton and Dr J. V. Watson, the Editors of this Series, for the opportunity to write this book and for their generous support and criticisms.

JOHN R. L. ALLEN
Spring, 1969

CONTENTS

CHAPTER 1

The Physical Background to Sedimentation

1.1 Physical System

Nearly everyone, at some time in his or her life, will have stood on a sea beach and watched how the waves repeatedly disturb and then deposit the sand grains, whilst the wind, blowing freely over the drier parts of the beach, gathers up some of the sand and sweeps it inland to form tall dunes.

This particular case, one from common experience, serves very well to introduce the general topic of this book, namely, the dynamic interactions, in natural surroundings, between detrital particles and moving fluids. These interactions, in which physical processes are of paramount importance, give us the depositional landscapes and seascapes with which we are familiar today and, at various removes in time, many of the sedimentary strata found in the earth's crust.

Various combinations of the three physical states of matter are involved in these interactions in the natural world. The *solid state*, represented by irregularly shaped and sized grains of mineral density, is common to all the interactions with a sedimentary product. But from common experience we know that forces must be exerted in order that these particles shall become shaped into morphological features and thereafter transformed into strata. These forces are of various origins, but chief amongst them are those directly resulting from the motion of *viscous fluids*, either the *liquid* we call water, as in the case of river and tidal flows, or the *gas* known as air, as in the case of the wind blowing over a ground or water surface. The driving forces causing the fluid motion can in turn be traced to the attraction of the earth for matter lying on or near its surface, or to the comparable attraction exerted at the earth's exterior by distant celestial bodies, particularly the sun and moon.

Turning from the large to the small, we know that matter consists of an association of molecules, each of which has a certain size depending on the amount of matter in the molecule. Now the strength and quality of the forces exerted by molecules on each other depend on the distance between the molecule centres. When in the case of simple molecules this distance is smaller that about 1×10^{-8}cm, the mutually exerted force is strong and repulsive. A weak attractive force is, however, exerted when the centres are further apart than about 5×10^{-8}cm. Stable equilibrium of one molecule relative to another is reached when the spacing between centres is of the order of $3\text{–}4 \times 10^{-8}$cm.

A solid is a coherent substance which strongly resists attempts to change its shape. In a solid the molecules are packed together about as closely as the short-range molecular forces will permit, their average spacing being of the order of the equilibrium spacing. Very commonly, the molecules in a solid have a spatially periodic, or crystalline, arrangement which is expressed as a virtually permanent molecular lattice. The molecules in the lattice oscillate about their mean stable positions but have negligible freedom to move about in the lattice from one position to another.

A liquid, on the other hand, is a substance which continues to deform readily when a small suitably chosen force is applied to it. Moreover, a liquid is a less coherent substance than a solid, for a portion of a liquid can be divided and recombined at will. These types of behaviour are related to the fact that in a liquid there exist no permanent regular arrangements of the molecules, although the spacing between molecule centres is still of the order of the equilibrium spacing. The spacing is, however, a little greater than in the corresponding solid, as one can see by comparing the densities of the same substance in the liquid and solid states, though this is not true of that exceptional and peculiar substance, water. Regular, or crystalline, arrangements of molecules do exist in liquids, but they are local and instantaneous in occurrence, building up and breaking down in a tiny fraction of a second. The molecules of a liquid are thus free to move about within the general confines of the substance. As is well known, there are some substances which are not obviously either solids or liquids. Pitch is one of these. It will shatter to pieces if struck a hammer blow, but will flow into all the corners of a containing vessel if left for a long enough period.

The distinction between liquids and gases is less fundamental than is that between fluids—liquids and gases together—and solids. Gases are, like liquids, readily deformed by the application of a suitably chosen force, though they lack even the modest degree of coherence shown by liquids. In a gas the molecules lie far apart, of the order of ten times the equilibrium spacing, and collisions between them are infrequent. Moreover, the large spacing means that the only effective intermolecular forces are the weak long-range attractive ones. In a gas, then, the molecules are exceedingly free to move from one part of the substance to another. Ordered arrangements of molecules are not attained.

1.2 Description of the System

In order to be able to see how the nature of detrital sediments depends on the properties of the environments in which they form, we need to be able to describe physically the system in which these deposits arise. Ideally, our system of description should be quantitative.

We begin by noting that we are chiefly concerned with the properties and behaviour of matter in the large, and not ordinarily with matter at the molecular scale. This does not absolve us from at first taking a small-scale view, for if we were able to look at a substance at the molecular scale, we would notice that the matter of which it consisted was concentrated at discrete places within the volume occupied by the substance, with large empty regions in between. We would also notice that the discrete particles of matter were in continuous random movement about mean positions and, if the substance were a liquid or gas, that random translation of the oscillating particles occurred.

In dealing with matter in the large, we make what is called the *continuum hypothesis*, by which we suppose that the physical properties of a substance are spread uniformly over the whole volume occupied by that substance and not, as in reality, concentrated in the molecules whose volume constitutes a tiny fraction of the whole. The continuum hypothesis may seem in the light of ordinary experience to be obvious and unnecessary, yet the whole basis of the mechanics of solids and fluids depends on it, for it permits us to consider elements of fluids or solids that are small compared to our smallest and most sensitive instruments yet very large compared to the sizes of molecules.

In the following chapters we shall be concerned with the continuum mechanics of natural systems in which granular solids are caused to flow from place to place by the action of fluid-applied or gravity-induced stresses.

Now mechanics is an exact science which depends on the measurement or prediction of physical quantities to the highest precision attainable. If we are to form an accurate understanding of the ways in which detrital sediments have come into being, we must attempt to specify the different aspects of the physical system in terms of numerical values of units of measurement. By and large, currents of water or air will not for us be either fast or slow, though the use of these vague terms is sometimes unavoidable, but will be said to flow at a number of units of length in a number of units of time. Intuitively we recognize that work is done when sediment is transported from one place to another by a stream of fluid. Now the notion of force is intimately bound up with the idea of rate of work—or power—and we can define force in one way as the product of mass and acceleration. Power, which is force multiplied by distance divided by time, is therefore quite a complicated quantity. Thus force and, by extension, work and power, are capable of precise definition. By using defined units of measurement we can make some attempt at describing in precise terms the physical systems we are interested in. Since the description is numerical and precise, we can exactly compare one system with another, provided consistent units of measurement are used. Such a method of description and comparison has obvious advantages over one based on purely qualitative statements.

The units of measurement used in mechanics are either fundamental or derived. The *fundamental units* of present interest are those of *mass* [**M**], *length* [**L**] and *time* [**T**]. The unit of mass is a measure of the amount of matter in a body, whilst that of length is a measure of the one-dimensional extent of the body in space. We can think of the unit of time as a measure of the duration of action of some process on a body, though the notion of time is not a simple one, as a little thought will show. The so-called *derived units* are all made by variously combining the fundamental units. Thus the unit of area is formed by multiplying so many units of length by so many units of length. As we have already seen, the units of velocity are obtained by dividing so many units of length by so many units of time.

It is most important in exact work to decide upon a system of units and, as far as possible, to adhere to that system. We shall use the centimetre-gram-second (CGS) system, wherein the centimetre (cm) is the unit of length, the gram (g) the unit of mass, and the second (s) the unit of time. Other systems, which will sometimes be found in books and papers, are based on the foot-pound-second (FPS) or the metre-kilogram-second (MKS).

Brief thought will show that the units used to describe physical quantities have *dimensions*, even those we have called fundamental. For example, if we keep in mind the idea of lines as capable of dividing up space into portions, the unit of length has a dimension of one in length and can be written $[\mathbf{L}] = [\mathbf{L}^1]$. Similarly, the unit of area has the dimensions of two in length and is $[\mathbf{L}\times\mathbf{L}] = [\mathbf{L}^2]$, whilst the unit of volume has the dimensions of three in length and becomes $[\mathbf{L}\times\mathbf{L}\times\mathbf{L}] = [\mathbf{L}^3]$. Since the velocity of a body is measured as distance travelled divided by duration of travel, the dimensions of velocity are one in length and minus one in time, i.e. velocity $= [\mathbf{L}/\mathbf{T}] = [\mathbf{L}\mathbf{T}^{-1}]$. The combinations $\mathbf{L}^3$ and $\mathbf{L}\mathbf{T}^{-1}$ are called the dimensional formulae of volume and velocity, respectively. In Appendix I will be found the dimensional formulae and names of the physical quantities used in this book. This table should be constantly referred to until the dimensional formulae of the quantities used have become thoroughly familiar.

It may be noted that some important physical quantities have dimensions of zero, that is, are dimensionless. One of these is angle, defined as arc divided by radius, which can be expressed as $[\mathbf{L}/\mathbf{L}] = [\mathbf{L}^0]$. Some dimensionless quantities are more complicated in structure, for example, the Froude number used in the fluid mechanics of flows with a free surface. Its definition and dimensions are

$$\frac{(\text{flow velocity})^2}{\begin{pmatrix}\text{flow}\\ \text{depth}\end{pmatrix}\times\begin{pmatrix}\text{acceleration}\\ \text{due to gravity}\end{pmatrix}} = \frac{\mathbf{L}^2\mathbf{T}^{-2}}{\mathbf{L}.\mathbf{L}\mathbf{T}^{-2}} = \mathbf{L}^0\mathbf{T}^0.$$

The specific gravity of a substance is also a dimensionless quantity, since it is defined as the ratio of two densities.

It is easy to see that the same dimensional formula can be shared by fundamentally different physical quantities. For example, stream depth and sand grain diameter both have the dimension of $\mathbf{L}^1$. A more complicated instance is that of the quantities known as the diffusivity of momentum (kinematic viscosity), the diffusivity of mass, and the diffusivity of heat, which all share the dimensional formula $\mathbf{L}^2\mathbf{T}^{-1}$.

The quantities of interest in mechanics can also be classified according to whether they are *vectors* or *scalars*. We distinguish vectors as those quantities which have the property of direction as well as magnitude. Velocity is an obvious and familiar vector quantity, but so can be area (it is possible to orientate an area in space), and pressure which is force directed perpendicularly over unit area. Scalar quantities which have magnitude alone can achieve directional significance only when mapped in space so that a gradient of the quantity is discernible. By mapping a scalar we obtain a new quantity which contains $\mathbf{L}^{-1}$ in its dimensional formula.

1.3 Representation of Fluid Motion

Some major aspects of the behaviour of moving fluids, or of bodies acted on by fluids, are *kinematic* and concerned alone with quantities, such as velocity, acceleration, or rate of discharge, which involve in their definitions only length and time.

If it is assumed, as is permissible, that water and air are incompressible in the context of natural sedimentary environments, we easily discover that a conservation law is operative when fluid flows from one place to another. We are here assuming in effect that the fluid density $[\mathbf{ML}^{-3}]$ remains constant throughout the region of flow. The conservation law can be illustrated very simply by considering the constant-discharge flow of water through a pipe, of small diameter in some places but of large diameter in others, which is completely filled by the water (Fig 1.1). If the rate of discharge of water from the end of the pipe is Q cm^3/s, then the rate of discharge through any other cross-section of the pipe must also equal Q, for if this were not so the density would have varied. Then

$$Q = Q_1 = Q_2 = Q_n = \text{constant}, \tag{1.1}$$

where Q_1, Q_2, Q_n are the discharges measured at arbitrary stations 1, 2, n along the pipe.

If we now imagine that the water emerges from the pipe as a coherent rod of constant cross-section equal to the cross-section of the end of the pipe, we easily see that so many units of length of water issue from the pipe in so many units of time. Hereby we are defining the velocity of flow from the end of the pipe. Now the notional rod of water that emerges from the pipe in unit time also has a certain volume, equal to the length of the rod times its cross-sectional area. But since the length of the rod was used to define the velocity of flow, it follows that

$$Q = UA, \tag{1.2}$$

where Q is the discharge, U is the mean velocity of flow from the end of the pipe averaged over the cross-sectional area of the end, and A is the cross-sectional area of the end of the pipe. The dimensional formula for fluid discharge is now easily seen to be $[L^2LT^{-1}] = [L^3T^{-1}]$. Eq. (1.2) is often referred to as the *equation of continuity* of an incompressible fluid. Returning to the pipe, since the discharge is the same for all cross-sections, it follows that

$$U_1A_1 = U_2A_2 = U_nA_n = \text{constant}, \tag{1.3}$$

where the subscripts denote different arbitrary stations. Hence a large cross-sectional area of the pipe means a small velocity of flow across the section, and *vice versa.*

Intuitively, it is unlikely that a small fluid element would follow a straight path through a pipe of varying cross-section such as is shown in Fig 1.1. Moreover, since all real fluids are viscous, we

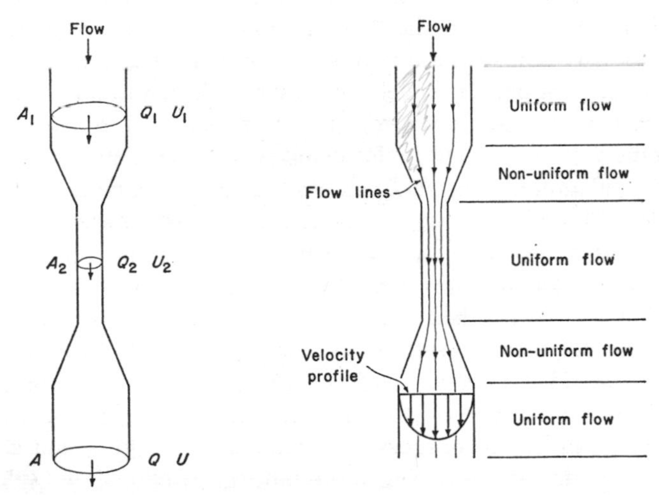

Fig 1.1 Fluid flow through a circular pipe of varying cross-sectional area.

would expect the speed of flow of different elements of the fluid to vary according to the distance of each element from the pipe wall. The question arises as to how in these circumstances we might depict the fluid motion and specify the velocity field over the region of motion.

Suppose that we measure the fluid velocity within the pipe of Fig 1.1 at a point fixed in relation to the pipe wall. Although different fluid elements move through the point, if the measured velocity remains constant with time, the flow is called *steady*. If, on the other hand, the velocity changed with time, because we varied the discharge through the pipe, we would then classify the flow as *unsteady*. But whether the flow is steady or unsteady depends on our choice of object with respect to which we take our chosen point as being fixed, i.e. our frame of reference. Consider the constant-velocity ascent of a balloon through the atmosphere. If our chosen point is fixed relative to the balloon, the motion of the air through the point does not change with time and the flow is steady. Now when the chosen point is fixed relative to the ground, the motion of the air is unsteady, for the velocity measured at the point changes as the balloon moves past the point.

By measuring the fluid velocity at many fixed points distributed over the region of flow, we can specify the whole *field of flow*. This specification will be particularly valuable if we make our observations at the same instant of time. In so doing we are taking what is known as the Eulerian view of the fluid motion, in which no attempt is made to determine the motion of every fluid element in detail. If the velocity at each point in a field of flow remains constant with time, then we have a *steady flow field*. The *flow field* is said to be *unsteady*, when the velocity at every point changes with time.

A consideration of Eq. (1.2) shows that the velocity measured on the axis of the pipe of Fig 1.1 changes with distance where the pipe is of varying cross-sectional area. Such a flow, in which the velocity measured at an instant varies from one point to another in the direction of flow, is called *non-uniform*. If the velocity were constant from one point to another along the direction of flow the fluid motion would be called *uniform*. A uniform flow would be encountered in our pipe only where, with the discharge held constant, the pipe was of uniform cross-sectional area. It follows that a flow which is both unsteady and non-uniform would show velocity variations both with distance and with time. Such a flow would be found in an expanding or contracting section of the pipe of Fig 1.1 at a time when the discharge through the pipe was varied. To describe a flow field as uniform or non-uniform is also to take an Eulerian view of the motion.

There is a second way of looking at fluid flow, known as the Lagrangian, in which we seek to determine in every detail the motion of each fluid element. This viewpoint presents a great many mathematical difficulties, but it has practical value, as it is fairly easy to mark a small element of fluid.

We can depict the fluid motion by plotting different kinds of *flow lines*. These can in many instances be calculated, but they can also be obtained experimentally, either from measurements of velocity or force or from studies of marked fluid elements and other tracers.

One of the most important flow lines is the *streamline*, defined as the imaginary line to which at any instant the velocity vector of all fluid elements lying on it is tangent. In uniform flow the magnitude and direction of the velocity vector remain constant over the entire length of any streamline. The streamlines of a uniform flow are therefore parallel to each other. In non-uniform flow, on the other hand, the streamlines are not parallel. It is easily seen that the magnitude of the velocity vector must increase as the streamlines converge but decrease where the streamlines spread further apart (Fig 1.1). When the flow field is steady the pattern of streamlines persists in time, since the magnitude and direction of the velocity vector at a point does not change. In an unsteady and non-uniform flow, on the other hand, a given pattern of streamlines can only have an instantaneous existence.

Streamlines not only have pattern but also numerical value. Their numerical value is that of their *stream function*, which is a constant for each streamline. To see what the stream function means we consider the motion of a fluid in a plane with the x-coordinate direction in the general line of motion and the y-coordinate direction at right angles to it. If u is the component of velocity of a fluid element parallel to the x-direction, and v the component parallel to the y-direction, the stream function ψ at the station occupied by the element is given by

$$u = \frac{\partial \psi}{\partial y}, \qquad v = -\frac{\partial \psi}{\partial x}. \tag{1.4}$$

The stream function is such that its derivatives yield the velocity components normal to the respective coordinate axes. It will be seen that the stream function has the dimensional formula $[L^2T^{-1}]$. The conservation law expressed by Eq. (1.2) is automatically satisfied by the stream function.

A streamline of great importance to the movement of sedimentary particles over the surface of the bed of a flow is that called a *limiting streamline* or *skin-friction line*. This streamline is obtained by considering the direction of the velocity vector at an infinitesimally small distance normal to the flow boundary. The velocity vector then becomes for all practical purposes parallel to the direction of action of the surface friction force (bed shear stress) exerted by the fluid. We have already seen that force is a vector quantity.

The two other types of flow line are streak lines and path lines. A *streak line* is the instantaneous locus of all fluid elements which have passed through the same point in the flow field. Streak lines are easily made, for example, by injecting dye or smoke into the flow from a small orifice. A *path line* is the trajectory, swept out during a definite period of time, of a suitably marked fluid element. Path lines are very often obtained experimentally by putting neutrally buoyant or exceedingly small particles into the flow, for example, balloons in air or plastic beads in water. Only in a steady flow are streamlines, streak lines, and path lines coincident.

1.4 Forces Acting on Fluids

We become involved with *dynamical quantities* as soon as the forces that act on fluids and solids in relative motion are considered. These quantities, of which density, momentum, force and energy are the most often encountered, all include the fundamental unit of mass in their dimensional formulae.

It is useful first to consider the general nature of the dynamical quantities of interest before we think of them in the context of fluids. From the definition already given, the dimensional formula for *force* is $[MLT^{-2}]$. Now we can regard a force as acting either at a point, in which case the dimensional formula for the quantity remains unchanged, or as having an intensity, i.e. as acting with a certain magnitude per unit of area, in which case the formula becomes $[ML^{-1}T^{-2}]$. For example, if a brick of mass m rests on one face on a horizontal table, the downward force acting on the brick as a whole is mg, where g is the acceleration due to gravity. However, if the area of the face of the brick in contact with the table is A, then the mean downward force acting per unit area of the contact face is mg/A. This last quantity is easily recognized as a pressure. Another important dynamical property is *momentum*. This is mass times velocity $[\mathbf{MLT}^{-1}]$, and can be thought of as a measure of the difficulty of stopping a body already in motion; *inertia*, with the dimensional formula $[\mathbf{M}]$, is a measure of the difficulty of setting in motion a body at rest, or of altering the motion of a body moving with a steady speed. In fluid mechanics it is often useful to talk about the momentum per unit volume $[\mathbf{ML}^{-2}\mathbf{T}^{-1}]$ of a fluid element. The *energy* of a body or fluid element can be measured in terms of its position relative to some standard position, or in terms of its motion. *Potential energy* is energy of the first kind and is measured by the work the body does in moving from its original position to the standard position, where the potential energy is considered equal to zero. Energy due to motion is called *kinetic energy*. For example, a body of mass m at a distance y above the surface of the earth does work equal to mgy in falling from rest back to the surface in the absence of retarding forces. But as $2gy = v^2$, where v is the final velocity of the body, the work done is $\frac{1}{2}mv^2$. Then

$$\text{energy} = mgy = \tfrac{1}{2}mv^2. \tag{1.5}$$

This is a general equation applicable to any situation like the above and is, moreover, an expression of energy conservation. It follows from Eq. (1.5) that the dimensional formula of energy is $[\mathbf{ML}^2\mathbf{T}^{-2}]$. Since we saw that the energy of the body was equal to the work done by it, the dimensional formula for work is identical with that for energy.

The next thing is to remind ourselves of the laws governing bodies in motion or at rest. These, due in their formal shape to Isaac Newton, are:

Law I: Every body continues in its state of rest or of uniform motion in a straight line unless impressed forces are acting on it.
Law II: Change of momentum per unit time is equal to the impressed force, and takes place in the direction of the straight line along which the force acts.
Law III: Action and reaction are always equal and opposite.

A little reflection on these laws will show that we can treat forces in terms either of energy change or momentum change.

The forces acting on fluid elements and on fluids and solids in relative motion are either *body forces* or *surface forces*. The body forces act on the whole bulk of a fluid element or of a solid immersed in a fluid. A surface force is one which acts at the surface of a body immersed in a fluid or over the surface of a fluid element.

Body forces act at a distance and depend on 'fields' analogous to the fields of force exerted by magnets or electrostatically charged objects. Thus we speak of a body that is falling towards the surface of the earth as acted on by the earth's gravitational field. We have already considered a number of cases of forces that arise through the action of the earth's gravity. A body moving on a curved path, for example, the tea or coffee in a cup that has just been stirred, can also be considered to exist in a force field, in this case a centrifugal acceleration directed outward from the centre of motion and equal to the square of the tangential velocity of a fluid element divided by the radius of curvature of the path of the element.

Surface forces act by direct contact between the surface of a body or fluid element and its surroundings. When the force is exerted perpendicularly to the surface of action, it is a normal force with a certain intensity, i.e. a *pressure*. When the force is exerted tangentially to the surface of action, it is a *shear stress* with again a certain intensity. The dimensional formulae of pressure and shear stress are therefore identical, but the one force is normal, the other tangential.

Fluids exert either static or dynamic pressures. We are well aware of the existence of *static pressures* in fluids at rest by the changes we experience in our ears during a dive into deep water or whilst climbing in a poorly pressurized aeroplane. Both experiences arise from the fact that in diving or ascending we vary the height, and thus the mass, of the column of fluid that exists vertically above us and is acted on by gravity. Consider an open vessel filled with water (Fig 1.2a). The pressure due to the atmospheric air exerted over an arbitrary element of the water surface of area A_1 is simply mg/A_1, where m is the mass of air existing in the vertical column above the arbitrary element. Let this pressure be p_A. Now consider a corresponding element of area A_2 at a depth y below the free surface of the water. The additional pressure acting on this element due to the mass of water is easily seen to be $p_W = \rho g y$, in which ρ is the fluid density. The total static (or hydrostatic) pressure acting on the element A_2 is therefore $p_A + p_W$. From this reasoning we see that the static pressure is proportional to the height of a column of fluid. The proportionality is explicit in the expression for the pressure exerted by the water, but is, of course, implied in that proposed for the atmosphere.

A fluid can exert a *dynamic pressure* only when it is in relative motion. We know this by having observed that the jet of water from a fireman's hose is capable of knocking down objects. Consider a free jet of water of density ρ, velocity U, and cross-sectional area A directed perpendicularly at a fixed wall (Fig 1.2b). Because the jet is normal to the wall, the momentum of the jet in the direction of the jet flow must be totally destroyed. But force is equal to the rate of change of momentum, so we deduce that the total force exerted by the jet on the wall is $\rho(AU)U$, where (AU) will be recognized as the discharge. The pressure exerted by the wall on the jet is therefore equal to ρU^2.

The equations of motion for a steady flow of fluid allow us to write for any point on a streamline

$$\tfrac{1}{2}\rho u^2 + p + \rho g y = \text{total energy} = \text{constant}, \tag{1.6}$$

in which ρ is the fluid density, u is the velocity of a fluid element at the point, p is the total fluid pressure at the point, and y is the elevation of the point relative to an arbitrary datum. This expression,

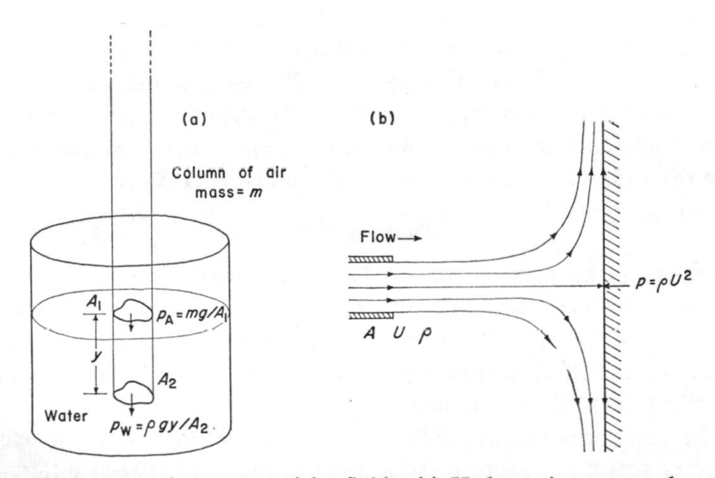

Fig 1.2 Pressure forces exerted by fluids. (a) Hydrostatic pressure due to a column of fluid. (b) Dynamic pressure due to a fluid in motion, in this case a free jet.

named after Daniel Bernoulli, is known as the *equation of energy*, for each of its terms has the quality and dimensions of energy per unit volume. It is a statement of the conservation, or interchangeability, of energy for flow along a streamline. The first term on the left will be recognized from Eq. (1.5) as the *kinetic energy per unit volume*; it is also known as the dynamic pressure, as can be seen from an examination of its dimensions. The third term on the left is obviously the *potential energy per unit volume.*

The Bernoulli equation is very powerful, with numerous applications. Consider the flow of a fluid around a cylinder whose axis is normal to the plane of the diagram (Fig 1.3). There is one streamline

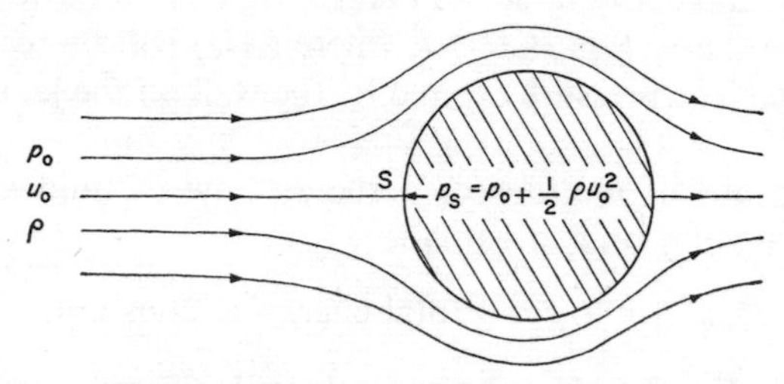

Fig 1.3 Fluid pressure exerted at an attachment (stagnation) point on an immersed circular cylinder.

which attaches to and divides on the cylinder at *S*. At *S* the component of the velocity of flow in the direction of the general stream is zero, and *S* is called a *stagnation* or *attachment point* of the stagnation streamline ending at *S*. If the velocity and pressure of the fluid at a distance upstream from the cylinder are u_0 and p_0, respectively, and the stagnation streamline is of constant elevation, then from Eq. (1.6) we can write for the pressure at *S*

$$\tfrac{1}{2}\rho u_0^2 + p_0 = p_s, \qquad (1.7)$$

where p_s is the pressure at *S*. The flow therefore exerts at *S* a dynamic pressure equal to $\frac{1}{2}\rho u_0^2$. We are in effect saying that there exists a force (pressure) gradient along the streamline dividing at *S* which has the effect of slowing down, or decelerating, fluid elements travelling along the streamline.

The other important surface forces are *shear stresses*. These exist in every real fluid where there is relative motion between different fluid elements or between the fluid and some solid object associated with the flow. Shear stresses arise because all real-world fluids are viscous, that is, they possess a property of 'thickness' or 'consistency' which represents an ability on the part of the fluid to resist deformation. We know from experience that much more effort has to be put into stirring glycerine than water, because glycerine is 'thicker', or more viscous, than water.

To see what *viscosity* means, consider a cube of foam rubber glued by one face to the top of a table (Fig 1.4). On the opposite face of the cube we glue a flat metal plate attached by a light string running over a pully to a weight pan. When the pan is unweighted, the piece of rubber is an undistorted cube. As more weights are added to the pan, however, the rubber distorts into an increasingly flatter rhombohedron. The degree of distortion is measured by the angle θ, whilst the force causing the distortion is clearly a *tangential* stress with an intensity mg/A, where m is the mass on the pan and A is the area of the upper face of the rubber cube. Now the distortion θ is said to be a *strain*. Clearly the relationship between the shear stress mg/A and the induced strain θ is going to be a definite physical property of the material of which the cube is made. We can define this property as the ratio of stress to strain. If, however, the cube of rubber were a fluid element, we would have to define the distortion as the rate of change with time of the strain measured by θ. Our fluid element will always tend to flow away from us, so we must keep on generating the distorting influences. Then

$$\text{viscosity} = \frac{\text{shear stress}}{\text{rate of change of } \theta \text{ with time}} \tag{1.8}$$

or, in symbols,

$$\mu = \frac{\tau}{du/dy}, \tag{1.9}$$

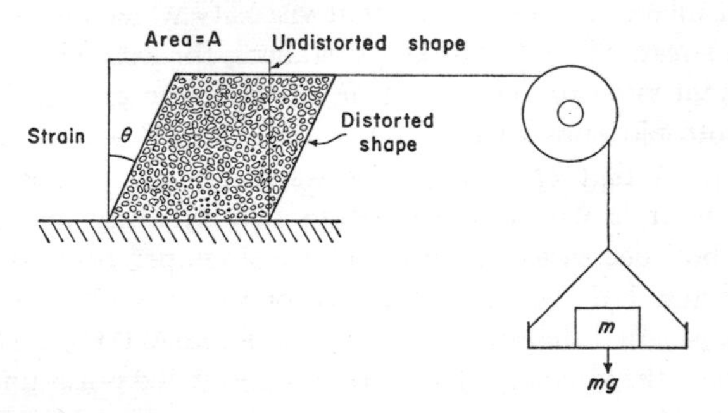

Fig 1.4 Distortion of a cube of foam rubber under an applied stress.

where μ is the viscosity $[ML^{-1}T^{-1}]$, τ is the shear stress, and du/dy is the velocity gradient measured normal to the direction of the fluid motion. The velocity gradient, which is the rate of change of θ with time for a fluid element, is the change of velocity u parallel to flow with distance y normal to flow. The velocity gradient has the dimensional formula $[T^{-1}]$.

The shearing behaviour of fluids is unfortunately not simple (Fig 1.5). Air and water, however, display a behaviour called

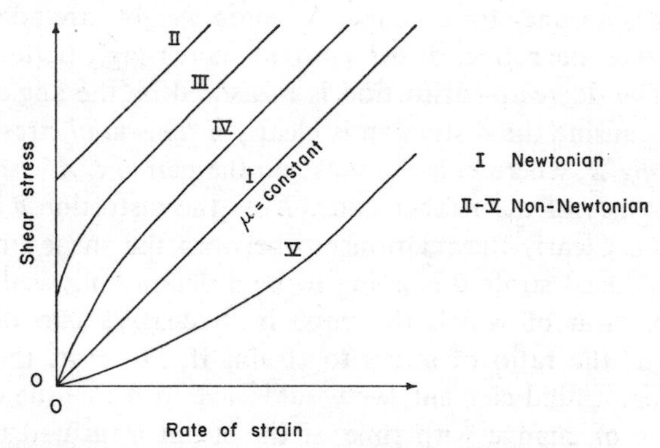

Fig 1.5 Stress-strain relationship for a Newtonian and various different non-Newtonian fluids.

Newtonian. Their viscosity is constant at a given temperature and pressure and is independent of the shearing stress or the duration of shearing. In the case of ideal, or Bingham, plastics, the viscosity is constant, but the substance will not flow until a certain stress value is exceeded. Other fluids, whose behaviour is pseudoplastic, visco-elastic or dilatant, have an apparent viscosity which depends on the shearing stress. Many fluid-like substances, for example, mud, have an apparent viscosity depending on the duration of shearing; these are thixotropic substances. The various fluids whose behaviour departs from that epitomized by water and air are called *non-Newtonian.* It is worth remembering that viscosity and apparent viscosity both decrease markedly with rise of temperature. At ordinary temperatures, however, the viscosity of water is about 0·015 dyn s/cm^2 (poise), and that of air at sea level about 0·000175 poise. For comparison, the viscosity of glycerine is about 300 poise under these conditions.

1.5 Laminar and Turbulent Flow

We know empirically of only two modes of fluid flow, *laminar* and *turbulent*. The nature of these was revealed in a particularly illuminating way by the English mathematician and physicist Osborne Reynolds. He made an experiment in which a narrow thread of dye was released at the centre of a long straight tube filled with steadily flowing water (Fig 1.6). When the flow velocity was below a certain

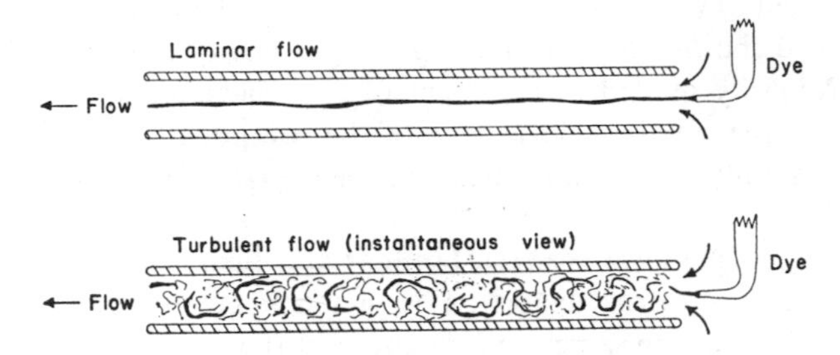

Fig 1.6 Reynolds' experiment on laminar and turbulent flow.

critical value, the dye thread was straight, coherent, and of practically constant width. This is laminar flow, characterized by the smooth linearity of the streamlines representing it (since the flow is steady, the streak line formed by the dye thread is also a streamline). When the flow velocity exceeded the critical value, however, the dye thread was at every instant found to be highly distorted, the distortion changing from instant to instant. Because the dye thread observed at an instant corresponds to a streamline, we can say that the instantaneous streamlines for the motion were also highly distorted. This is turbulent flow. Evidently, in turbulent flow, there is a transport of fluid in sizeable portions transversely to the mean direction of motion. Thus a turbulent flow comprises random secondary motions superimposed on a primary downstream translation.

It follows that, in a turbulent flow, the magnitude and direction of the velocity vector measured at a point must vary from instant to instant as different portions of the randomly eddying fluid pass the point. However, if we measure the velocity in a particular direction at the fixed point over long enough periods of time, we shall find that the measured velocity tends to a constant value.

Attaching orthogonal coordinates x, y and z to the fixed point, where x is parallel to the mean flow direction, we can write

$$\left.\begin{aligned} u &= \bar{u}+u' \\ v &= \bar{v}+v' \\ w &= \bar{w}+w' \end{aligned}\right\}, \tag{1.10}$$

wherein u, v and w are the velocities measured parallel to x, y and z, respectively. The bar denotes the velocity averaged over a long period of time (time-average velocity), and the prime the fluctuating component of velocity observed at any instant. The primed quantities are called the *turbulent fluctuating components of velocity*, and they are ordinarily quoted in the form of root-mean-square values (i.e. values of the standard deviations of the components). These values are commonly used as a measure of the degree of turbulence shown by a flow.

A major difference between laminar and turbulent flow is that a turbulent flow resists distortion to a much greater degree than a laminar flow of the same fluid. Because of the relative movement of sizeable portions of fluid in the turbulent flow, this flow would appear to have a very large viscosity compared to the laminar flow. The viscosity of the turbulent flow is, however, an *apparent* viscosity which varies with the character of the turbulence. We can write from Eq. (1.9) for the laminar flow

$$\tau = \mu \frac{du}{dy}, \tag{1.11}$$

but for the turbulent flow

$$\tau = (\mu+\eta)\frac{d\bar{u}}{dy}, \tag{1.12}$$

where η is the 'eddy viscosity'. Note the use of the time-average velocity in the last equation, Because η is ordinarily several orders of magnitude larger than μ, it follows that a turbulent flow exerts a larger shear stress than a laminar one for the same velocity gradient. The reason lies in the basic description of turbulent flow given in Eq. (1.10), for we can take the fluctuating velocity components in pairs and combine them with the fluid density to form a shear stress, for example,

$$\tau_{yx} = \rho\overline{u'v'}, \tag{1.13}$$

where τ_{yx} is the shear stress acting parallel to the direction of the mean flow, and the bar again denotes a time average. The time-average products of the fluctuating components differ from zero because as some fluid elements move upwards into regions of large time average, others must by continuity move downwards. Stresses such as τ_{yx} are called *turbulent stresses.*

Recalling Newton's second law, we see that the shear stresses given by Eqs. (1.11) and (1.12) are due to the transfer of momentum in a direction at right angles to the direction in which the stress acts. In the flows represented by these equations there is a continuous transport, or *diffusion,* of momentum from faster to slower moving levels in the flow. In laminar flows the transport occurs on a *molecular scale,* being due to the constant vibration and translation of the molecules. The transport occurs on a *macroscopic scale* in a turbulent flow, because of the transverse movement of sizeable parcels of fluid, the eddies of the turbulence. Then

$$\left.\begin{aligned} \tau_{\text{lam}} &= \frac{\mu}{\rho}\frac{d(\rho u)}{dy} \\ \tau_{\text{turb}} &= \frac{(\mu+\eta)}{\rho}\frac{d(\rho\bar{u})}{dy} \end{aligned}\right\}. \tag{1.14}$$

These equations say that the rate at which momentum is destroyed is proportional to the gradient of momentum, the proportionality factor in each case being the appropriate *diffusivity of momentum* (kinematic viscosity).

Analogously, we can write for any transportable fluid property, for dissolved substances, or for suspended particles

$$R = k_1 \frac{dc}{dy}, \tag{1.15}$$

where R is the transport rate of the property or matter parallel to y and c is the amount of the property or matter at a distance y from the source of what is transported. Eq. (1.15) is the *basic diffusion equation,* and it states, like Eq. (1.14), that the transport rate is proportional to the gradient of what is transported. The appropriate proportionality factor or diffusivity k_1 has the same dimensions of $[L^2T^{-1}]$ as the kinematic viscosity in Eq. (1.14). If the transport occurs in a moving fluid, we can write

$$\frac{(\mu+\eta)}{\rho}\frac{d(\rho\bar{u})}{dy} = k_1 \frac{dc}{dy}, \tag{1.16}$$

where Eq. (1.16) is called the Reynolds analogy. This equation, somewhat developed, is used to analyze the transport of fine sediment in turbulent suspension.

These additional reflections on turbulent flow illuminate still further the influences governing fluid motion. Evidently, *resistance to flow* can be either viscous, when it arises from the viscosity of the fluid, or inertial, when it arises from the masses of the sizeable portions of fluid engaged in the motion. However, in every fluid motion there must exist inertial as well as viscous forces, but it is the *balance between them* that significantly controls the character of the flow. This balance is succinctly expressed by the Reynolds number, defined as

$$\frac{\text{inertia force}}{\text{viscous force}} = \frac{UL\rho}{\mu} = \frac{UL}{\nu}, \tag{1.17}$$

in which U is a characteristic velocity of flow, L is a characteristic length for the flow system, and ν is the kinematic viscosity. The Reynolds number is without dimensions, as an examination of Eq. (1.17) will show. It is of fundamental importance.

As an illustration, there is a critical value of the Reynolds number at which the flow of a fluid in a tube changes from laminar to turbulent. This critical value, in the neighbourhood of 1×10^4, is independent of the tube diameter, or of fluid viscosity, provided that the other experimental conditions are unchanged. The importance of the Reynolds number therefore lies in the generality of its implications. The Reynolds numbers of most natural flows of fluid are found to be very large, whence it is hardly surprising that the flows are also turbulent.

1.6 Boundary Layers

When a fluid and a solid are in relative motion, a *velocity gradient* is set up at right angles to the direction of flow, because of the viscous and other retarding forces acting within the fluid. The fluid in contact with the surface of the solid is brought to rest relative to the solid, and there is said to exist a condition of 'no slip' between fluid and solid. It follows from Eqs. (1.11) and (1.12) that there exists a shear stress at the solid surface which opposes the motion of the fluid past the solid, or of the solid past the fluid, whichever is the case. There will also be shear stresses in those parts of the fluid where the velocity gradient exists. But as the velocity gradient diminishes in magnitude with increasing distance normal to the solid surface, the shear stresses in the fluid must also diminish in the same direction, by Eqs. (1.11) and (1.12). At a certain distance from the surface, the velocity gradient for all practical purposes becomes zero, and so no internal stresses need be considered to exist at this and larger distances. Hence there exists a region of the flow in contact with the solid surface which is affected by retarding stresses, and a more distant region of the flow which is unaffected by the presence of the solid surface. The region of stress, or velocity gradient, is called the *boundary layer* of the flow, whereas the unstressed region is called the *free* or *external stream.*

The study of boundary layers, initiated by Ludwig Prandtl in the early decades of the present century, is of very great practical importance.

If a smooth flat plate is placed parallel to flow in a fluid stream of uniform velocity U_∞, a boundary layer will be found to develop on the plate (Fig 1.7). The boundary layer thickens up gradually downstream from the leading edge, and over a certain distance from the edge is laminar. The equation for the velocity profile of the laminar boundary layer measured at right angles to the flow direction is

$$\frac{u}{U_\infty} = \frac{3}{2}\left(\frac{y}{\delta}\right) - \frac{1}{2}\left(\frac{y}{\delta}\right)^3 \quad (y \leqslant \delta), \tag{1.18}$$

in which u is the velocity parallel to the x-direction at any distance y normal to the plate and δ is the boundary layer thickness where

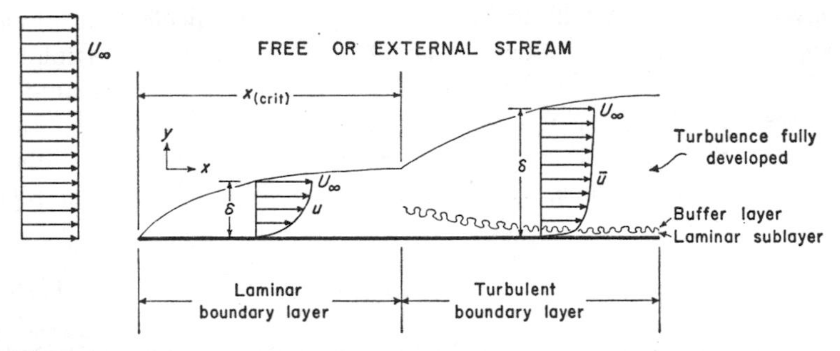

Fig 1.7 Development and structure of a boundary layer on a thin flat plate immersed in a parallel uniform stream.

the profile is measured. When $y > \delta$, the velocity u is equal to the velocity U_∞ of the external stream. The thickness of the boundary layer is given by

$$\delta = 4{\cdot}64\left(\frac{\mu x}{\rho U_\infty}\right)^{\frac{1}{2}}, \tag{1.19}$$

where x is the distance from the leading edge of the plate. These equations state that the velocity profile of the laminar boundary layer is approximately parabolic, and that the layer thickens according to the square root of the distance from the start of the layer.

Beyond a certain distance $x_{(crit)}$ on the plate, however, the inertial forces acting in the boundary layer become so large relative to the viscous ones that turbulence sets in. The Reynolds number for the onset of turbulence is of the order of 1×10^5, when based on $x_{(crit)}$. The boundary layer remains turbulent beyond this distance.

Turbulent boundary layers are difficult to analyse because of the non-uniqueness of the apparent viscosity of turbulent flow. However, for the turbulent boundary layer on the plate, we can write

$$\frac{u}{U_\infty} = \left(\frac{y}{\delta}\right)^{\frac{1}{7}} \quad (y \leqslant \delta), \tag{1.20}$$

for the velocity profile, in which δ is the thickness of the layer where the profile is measured. The layer thickens as

$$\delta = 0{\cdot}376x\left(\frac{\mu}{\rho U_\infty x}\right)^{\frac{1}{5}}. \tag{1.21}$$

Eq. (1.20) is often called Prandtl's one-seventh power law.

Provided the plate and fluid were extensive enough, the boundary layer of Fig 1.7 would go on thickening until it affected the whole flow field. When a boundary layer fills the whole flow field available to it, the layer is said to be *fully developed.* The flows encountered in pipes and channels, and on natural surfaces, are ordinarily fully developed boundary layer flows. In the case of a fully developed laminar flow in a pipe, we can write for the shear stress on the pipe wall

$$\tau_0 = \frac{3}{2}\mu\frac{U}{r}, \tag{1.22}$$

where τ_0 is the stress at the wall (the boundary shear stress), U is the flow velocity at the centre of the pipe, and r is the pipe radius.

In the case of the fully developed turbulent boundary layer, we have for the boundary shear stress

$$\tau_0 = 0{\cdot}0228\rho U^2 \left(\frac{\mu}{\rho U r}\right)^{\frac{1}{4}}, \tag{1.23}$$

in which U is the flow velocity at the centre of the pipe and r is the pipe radius.

The turbulent boundary layer is divisible vertically into three distinct layers differing in the microstructure of the flow (Fig 1.7). The lowest layer, in contact with the solid boundary, is called the *laminar sub-layer*. In this layer, which is very thin compared to the whole boundary layer, viscous forces predominate over inertial ones and there is no turbulent eddying, although there are random spiral vortical motions with axes parallel to flow. The velocity gradient in the laminar sublayer is very nearly linear with distance from the solid surface. Next outward from the laminar sublayer is the *buffer layer*, where the flow is neither fully laminar nor fully turbulent. This layer is several times thicker than the laminar sublayer but is still thin compared to the whole boundary layer. Beyond the buffer layer is the *fully turbulent* part of the boundary layer.

A consideration of the structure of turbulent boundary layers has led to the development of the *universal velocity profile law* of turbulent flow, due in large measure to Theodore von Kármán. From experimental tests of the theory, we can write for the velocity profile of any turbulent flow above any smooth boundary

$$\frac{u}{\sqrt{(\tau_0/\rho)}} = 5{\cdot}5+5{\cdot}75\log_{10}\left(\frac{y}{\nu}\sqrt{\frac{\tau_0}{\rho}}\right), \tag{1.24}$$

and for any rough boundary

$$\frac{u}{\sqrt{(\tau_0/\rho)}} = 8{\cdot}48+5{\cdot}75\log_{10}\left(\frac{y}{K}\right), \tag{1.25}$$

in which u is the velocity at a distance y normal to the boundary and K is a length describing the size of the roughness elements (sand grains, cavities, etc.) on the flow boundary. Note that the quantity $\sqrt{(\tau_0/\rho)}$, involving the boundary shear stress, has the dimensions of a velocity. It is the *shear velocity*, usually given the symbol U_*. Eqs. (1.24) and (1.25) are of enormous practical value, as they permit boundary shear stress to be calculated from measured velocity profiles, and *vice versa.*

1.7 Separation of Flow

Separation of flow occurs where a boundary layer detaches from a solid surface and enters the main body of fluid as a *free shear layer*. Very commonly, the free shear layer reattaches to the solid boundary at some point downstream from the separation point. Flow separation ordinarily leads to the occurrence in the flow of variously orientated 'rollers' or 'vortices' which go under the general title of *separation bubbles*. Separated flows are of great importance in the study of many sedimentary structures.

To illustrate the nature of flow separation we will describe an experiment in which we study the flow over a low barrier built transversely across the bed of a flume channel (Fig 1.8a). We can traverse a dye thread in the stream to find out the pattern of motion. Upstream of the barrier the motion of the dye reveals a transverse zone of sluggishly recirculating fluid over which the external stream rises on a curved path. The point *S*, where a streamline detaches from the flow boundary, is called a *separation point*, and the point *A*, where the streamline reattaches to the bed, an *attachment point*. This streamline, known as a *separation streamline*, divides off the recirculating fluid from the external flow. On the downstream side of the barrier, we find another transverse zone of sluggishly recirculating fluid, and we can again draw a streamline that detaches from and returns to the flow boundary. Note that the streamlines of both

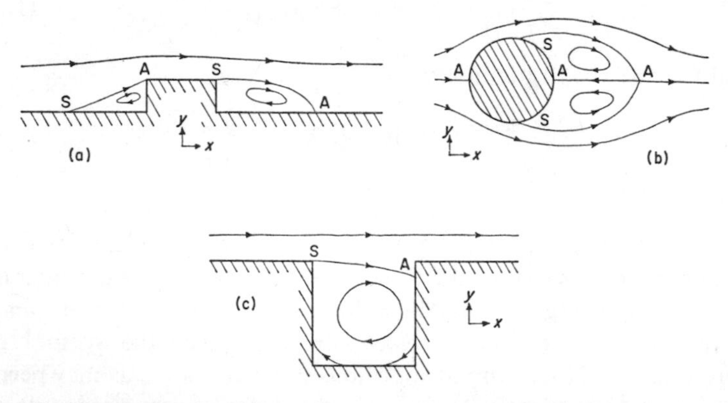

Fig 1.8 Flow separation (a) at a low transverse barrier. (b) downstream of an immersed sphere. (c) at a transverse slot.

zones of recirculation are closed loops. Note also that there are strong velocity gradients in the neighbourhood of the separation streamlines. These gradients denote the layers of free shearing in the flow.

We could have obtained a similar result by studying the low-velocity flow of water around a large smooth sphere (Fig 1.8b), when a ring-shaped separation bubble would have been detected. In this case there is no zone of recirculation in front of the body disturbing the flow. We also note that attachment of the separation streamline occurs in the body of the flow.

Instead of the barrier of Fig 1.8a, we could have used a slot of square section arranged transversely across the flume channel. A recirculating separated flow would have been found to fill the slot (Fig 1.8c).

These experiments, combined with a study of Eqs. (1.6) and (1.7), show that flow separation occurs whenever a fluid stream of suitably large Reynolds number encounters a sufficiently adverse pressure gradient or expands quickly over a sharp edge. Separation can occur in the lee of bluff bodies placed in a stream and, under appropriate conditions, in front of bluff objects. Flow separation will not occur, however, when a body is streamlined relative to a moving fluid. There are very few natural flows whose Reynolds numbers are so small that flow separation would not occur in them. We shall see that such common sedimentary structures as ripples, dunes, and flute marks are all associated with separated flows to leeward. The U-shaped scours formed around pebbles and boulders in stream beds or on beaches also depend on flow separation, as do sand drifts in the lee of obstacles.

The flow fields of Fig 1.8a, c are *two-dimensional*, meaning that the character of the flow field would be substantially unchanged if we took other planes of view parallel to the (x, y) plane. Most natural separated flows are *three-dimensional*, changing in character in all directions, often because the objects which give rise to them have a limited extent transversely to flow. Three-dimensionality of the separated flow can also arise if the body causing separation is of uneven height or depth across the flow, or has an edge which is not at right angles to flow.

Because of these factors it is useful to distinguish two varieties of separation bubble, *rollers* and *vortices* (Fig 1.9). Rollers arise when the discontinuity causing separation is skewed between 45° and 90°

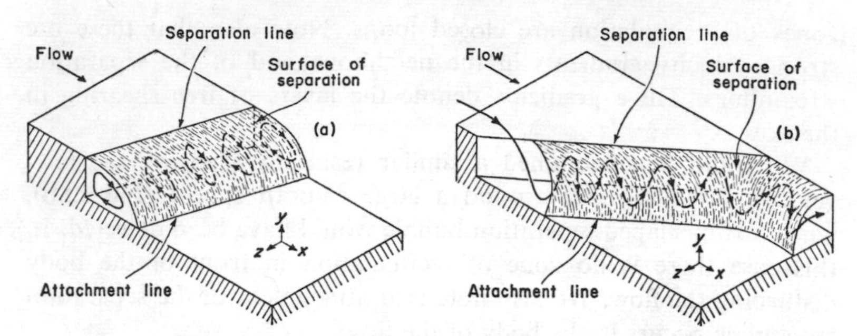

Fig 1.9 Three-dimensional separated flows. (a) Roller. (b) Vortex.

to the flow; the streamlines of a roller are closed loops, and the surface of separation isolates the roller completely from the external stream above. A vortex, on the other hand, arises where the discontinuity causing separation is skewed between 0° and 45° from the direction of flow. In a vortex the streamlines are helical spirals, denoting a mass movement of fluid down the axis of the vortex. Consequently, the surface of separation bounding the vortex must have a 'hole' in it somewhere upstream, to let the stream in, and a second 'hole' somewhere downstream, to let the flow out, so that continuity can be satisfied. There is a variety of separated flow called a secondary flow. One type of secondary flow consists of one or more pairs of oppositely rotating helical spiral vortices with axes parallel to the flow. Another type of secondary flow, found particularly in curved channels, consists of a single helical spiral vortex.

Obviously the flow fields associated with three-dimensional separated flows are complicated. From the geological standpoint, however, it is the patterns of skin-friction lines associated with such flows that are most interesting. For sand grains would roll, and directional structures would align themselves, parallel to skin-friction lines. We find that the patterns of skin-friction lines associated with

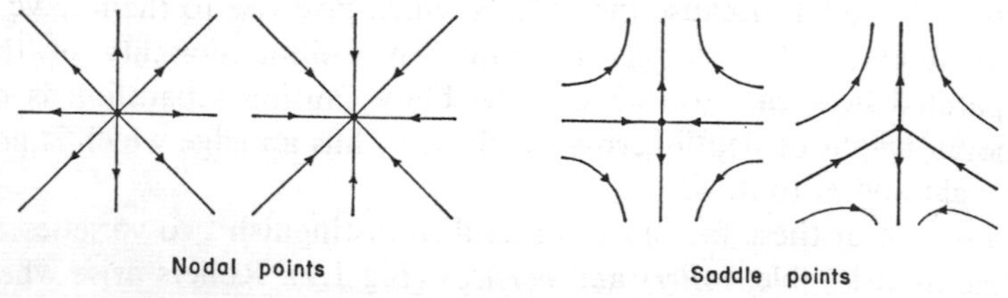

Fig 1.10 Singularities in fields of skin-friction lines.

natural separated flows consists of networks of lines that join or branch at only two kinds of point, or singularity (Fig. 1.10). One of these is called a *node*, the other a *saddle*. Nodes and saddles can be either points of separation or of attachment, depending on the directions of the shear-stress vector. Obviously where a nodal attachment point occured on a sand bed, the grains would be rolled away in all directions and a hollow would arise on the bed. Where there was a saddle point, some kind of linear ridge might be expected to form by a combination of erosion and deposition.

1.8 Turbulence in Boundary Layers and Separated Flows

We have now seen that two different kinds of shear layer—or zone of velocity gradient—can exist in fluid streams. One of these kinds is a *boundary layer* in contact with a solid surface and owing its existence to the relative motion between fluid and solid. The other kind, the *free shear layer*, is found within the body of a flow, and clearly depends more on the relative motion between two streams of fluid than on the presence in the flow of a solid surface. Like a boundary layer, a free shear layer can be either laminar or turbulent, according to the size of the Reynolds number. However, almost all natural free shear layers are turbulent, and so we shall not discuss further the laminar layers of this sort. The importance of turbulent free shear layers lies in the larger amounts of work of certain kinds that they can do relative to turbulent boundary layers.

In a turbulent boundary layer, shear stresses such as τ_{yx} of Eq. (1.13) vary in magnitude with increasing distance normal to the boundary, as is shown by the plot of τ_{yx} itself in Fig 1.11. The intensity of the stress is, of course, practically zero within the laminar sublayer, but outwards through the buffer layer the intensity rises to a maximum close to the outer edge of the layer. The stress gradually decreases in intensity through the fully developed turbulent part of the boundary layer. Where the maximum stress is reached, close to the edge of the buffer layer, the magnitude of τ_{yx} is almost equal to that of the boundary shear stress, the difference being the viscous element added in the laminar sublayer. That τ_{yx} should vary across the flow is not, of course, surprising, since we saw that the apparent viscosity of a turbulent flow is a function of the turbulence intensity. A convenient way of relating the magnitude of the turbulent

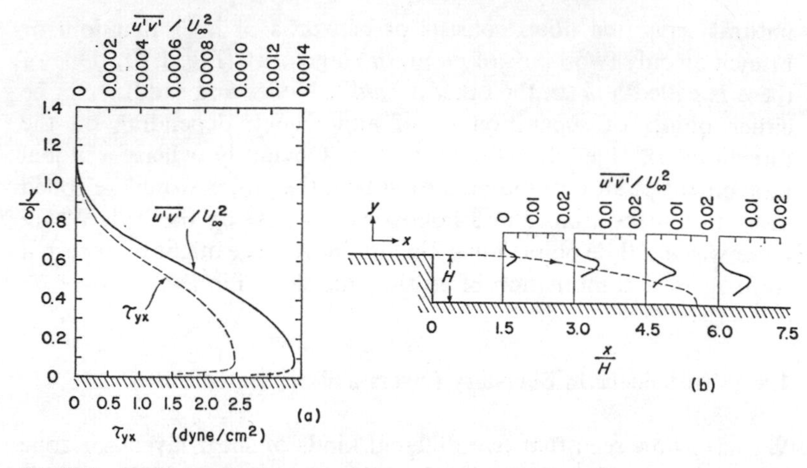

Fig 1.11 Spatial variation of turbulent stress and velocity products. (a) Shear flow past a wall (data of P. S. Klebanoff). (b) Free shear flow at a downward step (data of I. Tani).

fluctuations at their maximum to the average character of the flow is to compare the product $\sqrt{(\overline{u'v'})} = \sqrt{(\tau_{yx}/\rho)}$ to the velocity of the free stream U_∞. Experiment has shown that for a smooth boundary layer flow the maximum value of the ratio $\sqrt{(\tau_{yx}/\rho)}/U_\infty$ is of the order of 0·04.

Turning now to free shear layers, we must remember that these are boundary layers which have become detached from a solid surface. Nevertheless, they have a character which is quite different from that of true boundary layers. For one thing, free shear layers are markedly unstable, breaking down within a small distance from separation into powerful vortices. These, as they are convected down the layer, rapidly grow in size and vigour, but eventually decay into random turbulence. Thus when a turbulent boundary layer separates, the resulting free shear layer is more turbulent than the parent. As shown in Fig 1.11, the ratio $\sqrt{(\tau_{yx}/\rho)}/U_\infty$ in a turbulent free shear layer reaches a maximum value typically between 0·08 and 0·15, twice as large as that in the corresponding boundary layer. The heightened turbulence, where it comes into contact with the bed in the neighbourhood of an attachment point, can exert very large instantaneous forces on the bed. In a natural situation, this could mean an increased agitation and transport of sand, or a heightened rate of erosion of mud.

1.9 Fluid Drag

The concept of fluid drag is simply the recognition that when a fluid and a solid body are in relative motion, there arise forces which oppose the motion and enforce equilibrium. We met the concept in one context in the discussion of boundary layers, but will now view it from a less particular standpoint.

There are three types of drag. *Viscous drag* occurs at very small Reynolds numbers, when viscous forces predominate over inertial ones, because of deformation of the fluid not merely close to the surface of the solid with which the fluid is in relative motion, but also over a very considerable distance away from it. With gradual increase of the Reynolds number, viscous drag gives place to *surface drag* as the layer of rapid velocity change becomes thinner and thinner and pressed more and more closely against the surface of the body. The third kind of drag, known as *form drag*, is associated with flow separation at discontinuities of the shape of the body. Unless the body is streamlined, form drag will always accompany surface drag and may predominate over surface drag. Form drag is important geologically in the flow of water over beds with current ripples, dunes, or flute marks.

From energy considerations, we can write the basic *drag equation* as

$$F_D = C_D A\rho \frac{U^2}{2}, \tag{1.26}$$

in which F_D is the drag force opposing motion, A is the projected area of the solid body on a plane normal to the motion, ρ is the fluid density, and U is the relative velocity of fluid and body. C_D is a variable dimensionless coefficient of drag given by

$$C_D = \frac{F_D}{\frac{1}{2}A\rho U^2}, \tag{1.27}$$

and is a function of Reynolds number and body geometry. These equations can be applied to flow in pipes and channels, to the flow of fluids past stationary immersed bodies, and to the fall or ascent of bodies in still fluids.

Eq. (1.26) has been solved without recourse to experiment in the important case of the settling of single spheres, assumed smooth and solid, in still fluids of wide extent. Here the driving force is a body force equal to the volume of fluid displaced by the sphere times the product of the density difference between fluid and solid and the acceleration due to gravity. Equating the driving force with the resisting force arrived at mathematically, we have

$$\frac{4}{3}\pi\left(\frac{D}{2}\right)^3(\sigma-\rho)g = 6\pi\left(\frac{D}{2}\right)\mu V_0, \tag{1.28}$$

in which D is the diameter of the sphere, σ is the density of the sphere, ρ is the fluid density, and V_0 is the free falling velocity of the sphere. Solving for V_0 we obtain

$$V_0 = \frac{1}{18}\frac{(\sigma-\rho)gD^2}{\mu}, \tag{1.29}$$

which is *Stokes law of settling*. But from Eq. (1.27) we can write

$$\frac{4}{3}\pi\left(\frac{D}{2}\right)^3(\sigma-\rho)g = 6\pi\left(\frac{D}{2}\right)\mu V_0 = \frac{24}{V_0 D\rho/\mu}A\rho\frac{V_0{}^2}{2}, \tag{1.30}$$

whence we find that

$$C_D = \frac{24}{V_0 D\rho/\mu} = \frac{24}{\left(\begin{matrix}\text{Reynolds number}\\ \text{of particle}\end{matrix}\right)}. \tag{1.31}$$

Experiment has shown that Eq. (1.31) accurately predicts the free falling velocity of a sphere up to a Reynolds number of about unity (Fig 1.12). At larger Reynolds numbers, effects due to inertia and form drag become increasingly important and the value of C_D declines more and more gradually with ascending Reynolds number. At large Reynolds numbers the free falling velocity of a sphere is approximately proportional to the square root of the sphere diameter.

The above treatment rests on the supposition that we are dealing with a single sphere in a medium of wide extent compared to the sphere. It is much more realistic to focus attention on one sphere which is settling in a medium containing other settling spheres. The effect of the other spheres is to reduce the falling velocity of the sphere under consideration. The following equation has been found to apply when the spheres are all of one size:

$$V_s = V_0(1-C)^n, \tag{1.32}$$

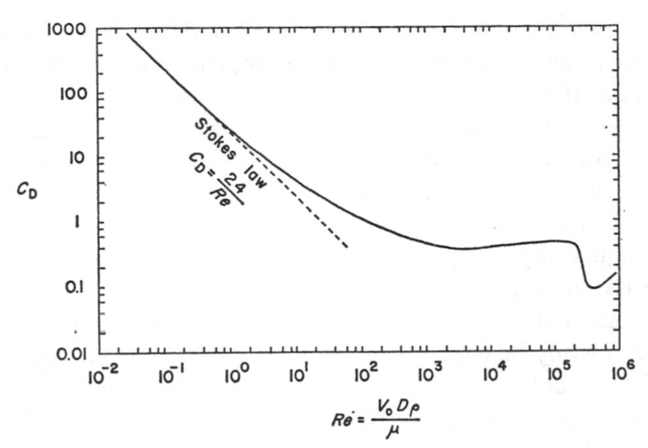

Fig 1.12 Experimental curve of drag coefficient of a sphere as a function of Reynolds number.

wherein V_s is the falling velocity in the suspension, V_0 is the free falling velocity, C is the volume concentration of spheres in the suspension (equal to occupied space/total space), and n is an exponent varying between 2·4 in the region of Stokes law and 4·6 where the Reynolds number is large. As can be seen, the effect of concentration is substantial, particularly at large particle Reynolds numbers.

We now turn from a consideration of drag when the body is completely immersed in the fluid, to drag when a body encloses a moving stream, as illustrated by flow through pipes of circular cross section. For this problem Eq. (1.26) is rewritten as

$$\tau_0 = \tfrac{1}{8} f \rho U^2, \tag{1.33}$$

in which f is a dimensionless drag coefficient known as the Darcy-Weisbach coefficient. Experiment has shown that the value of f, like that of C_D for the sphere, depends on Reynolds number and the geometry of the body involved in the relative motion, in this instance the pipe walls. When the pipe walls are perfectly smooth, we obtain experimentally the two linear plots shown in Fig 1.13, the steeper line corresponding to laminar flow and the less steep one to the turbulent regime. Now when the pipe walls are roughened by glueing sand grains to them so as to form a continuous cover, the value assumed by the Darcy-Weisbach coefficient is found to vary with the relative roughness, as measured by the ratio of the pipe radius r to the roughness dimension K as represented by the diameter of the sand grains. It can be seen from Fig 1.13 that we obtain a different curve of f against the Reynolds number depending on the relative roughness. If we choose a constant Reynolds number, the value of f increases as K approaches closer to r in value, namely, with ascending relative roughness. If we choose a constant relative roughness, however, we see that at sufficiently large Reynolds numbers the value of f becomes independent of Reynolds number. This means that at these sufficiently high Reynolds numbers, viscous influences no longer control the flow.

The essential meaning of these somewhat complicated relationships can easily be deduced by referring to Eq. (1.33). Evidently, if we wish to obtain flows of constant Reynolds number, we must exert a larger driving force when a pipe is rough than when it is smooth, because the rough pipe presents the greater resistance. But since for a steady flow the resisting force is equal and opposite to the driving force, we find that the flow through a rough pipe exerts a larger boundary shear stress than that through a smooth pipe, the Reynolds numbers being supposed equal.

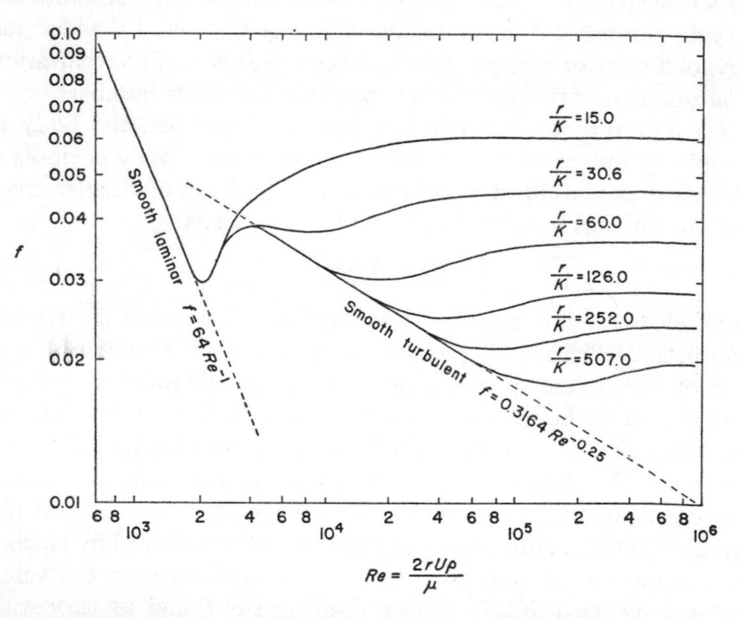

Fig 1.13 Experimental curves of the Darcy-Weisbach coefficient as a function of Reynolds number and relative roughness for flow through circular pipes (data of J. Nikuradse).

Eq. (1.33) given above for flow through pipes can be applied after only minor modification to flow through open channels, including those of rivers and tidal streams.

1.10 Entrainment of Sediment

Experiments by numerous workers have shown that as the velocity of fluid flow over a bed of sediment is increased, there comes a stage when the intensity of the applied force is large enough to cause sediment to move from the bed into the flow. This stage is variously known as the *threshold of movement*, or the critical stage for erosion or entrainment. It can be specified in terms of a value of the boundary shear stress, the *critical boundary shear stress.*

The value of the critical boundary shear stress depends chiefly on the nature of the sediment making up the bed, which may be distinguished as either *cohesionless* or *cohesive*. Cohesionless sediments are formed of loose grains which are not bound together in any way by surface or electrochemical forces. Clean sands and gravels are the commonest cohesionless sediments. When they are eroded, particles leave the bed singly, and not in clusters bonded together. As we shall see, the critical stress for cohesionless sediments can be arrived at analytically. Cohesive sediments, on the other hand, consist of grains that are bonded together more or less strongly by surface and electrochemical forces. Muds formed of silt or clay sized particles are the commonest cohesive sediments, and the material entrained from such beds generally consists of variously sized clusters of clay or silt particles and practically never of single flakes. Moreover, experiment has shown that the average size of the clusters entrained depends not only on the bulk properties of the bed but also on the character of the entraining flow. Thus we are at present unable to treat analytically the entrainment of cohesive materials.

In order to see what controls the entrainment from a cohesionless bed, consider the equilibrium beneath a steady fluid stream of a spherical particle of diameter D and density σ resting on a bed of similar particles (Fig 1.14). The stress at the threshold will be given by equating the fluid forces that just cause movement of the grain to the body force holding the grain in place. The body force is easily seen to be

$$F_g = \tfrac{1}{6}\pi(\sigma-\rho)gD^3. \qquad (1.34)$$

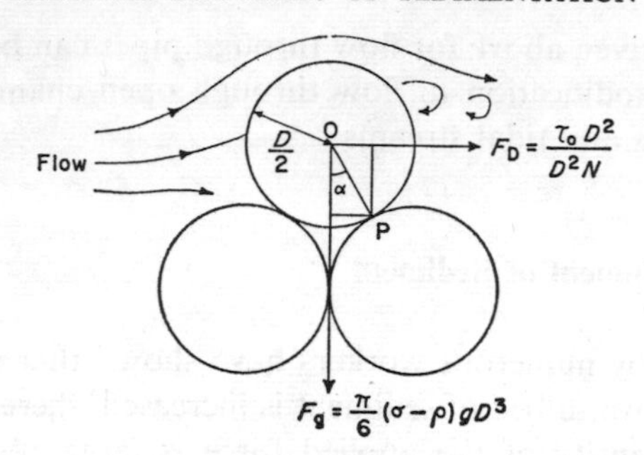

Fig 1.14 Equilibrium of a spherical particle on a bed of similar particles beneath a fluid stream.

Now the fluid stream exerts on the bed a tangential drag force per unit area of τ_0, whence the fluid force acting on the single grain becomes

$$F_D = \frac{\tau_0 D^2}{D^2 N}, \tag{1.35}$$

in which N is the number of grains exposed to the drag on unit area of the bed. The grain in moving off the bed will pivot about a point P in Fig 1.14, and the threshold of movement will be reached when the moment of the drag force about the pivot equals the moment of the body force about the pivot. Then

$$\tau_{(crit)} = \tfrac{1}{6}\pi D^2 N \tan\alpha\,(\sigma-\rho)gD, \tag{1.36}$$

in which $\tau_{(crit)}$ is the critical stress, the angle α is related to the packing of the grains, and the quantities to the left of the bracket can be regarded as forming a complex constant. In terms of the above analysis, then, the critical stress is directly proportional to the particle diameter (D^2 and N have a reciprocal relationship in the part of Eq. (1.36) forming the complex constant).

Instead of using the critical boundary shear stress alone to define the threshold of movement, it is often convenient to employ a *dimensionless form of the stress.* We can write

$$\theta_{(crit)} = \frac{\tau_{(crit)}}{(\sigma-\rho)gD}, \tag{1.37}$$

in which $\theta_{(crit)}$ is a dimensionless threshold stress criterion. Fig 1.15 shows an experimental plot of $\theta_{(crit)}$ against D for quartz-density grains in water. It will be seen that the relationship between the threshold stress and particle size of cohesionless material is not as simple as suggested by the preceding analysis. This is because the character of the flow round the grains varies with the grain size and the velocities of flow near to the grains. Quartz-density grains in water obey Eq. (1.36) only for $D \geqslant 0{\cdot}6$ cm, when $\theta_{(crit)}$ in Fig 1.15 attains a constant value of about 0.06.

It is unfortunate that we have such a limited physical understanding of the entrainment and erosion of cohesive beds, if only because many important sedimentary structures arise through current action on mud. The position is further complicated by the fact that a mud bed can be eroded in two distinct ways. Either the fluid stresses by their direct action cause lumps to be detached from the bed, or the stresses act indirectly by way of the impact with the bed of grains being carried in the flow, in a manner analogous to sand blasting. The two modes of erosion may very often occur together.

When the fluid stream by its direct action tears lumps off a mud bed, we can write for the critical boundary shear stress

$$\tau_{(crit)} = k_2 \tau_{mud}, \tag{1.38}$$

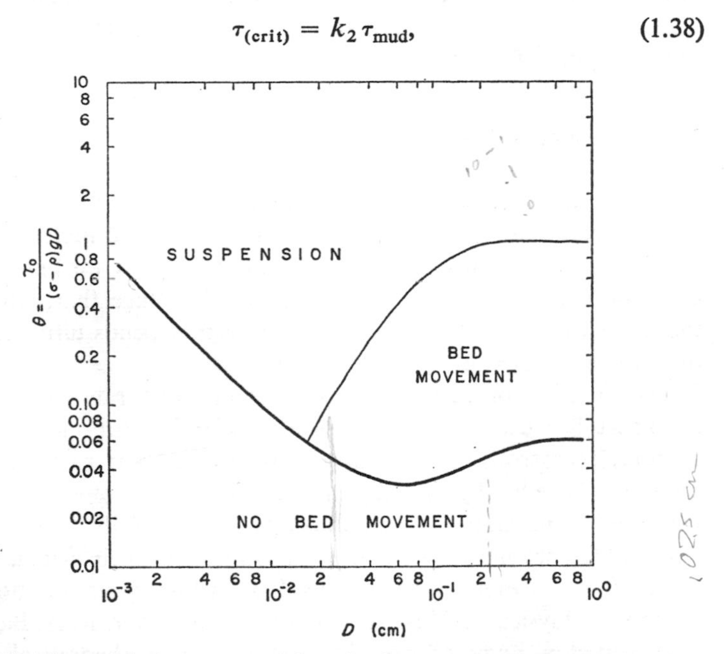

Fig 1.15 Threshold of movement and mode of transport of quartz-density solids in water (after **A. Shields** and **R. A. Bagnold**).

in which τ_{mud} is the bulk shear strength of the mud bed, and k_2 is a variable factor taking into account the character of the flow and the bed surface. The rate of erosion of the bed, measured as mass removed from unit area in unit time, then becomes

$$E = k_3(\tau_0 - \tau_{(crit)})^n, \tag{1.39}$$

in which E is the rate of erosion, and k_3 is a proportionality constant, and, from experimental data, $n \approx 2$. If the stresses act indirectly, however, we must write

$$E = Nm, \tag{1.40}$$

where N is the number of contacts made by the transported grains with the bed in unit time and area, and m is the average mass of bed material removed at each contact. Both N and m will increase with ascending turbulence of the flow, and so we can infer that the erosion rate given by Eq. (1.40) varies as some power greater than unity of the boundary shear stress. However, the erosion of mud beds as the result of the impact of grains in transport over the bed, could occur at much lower fluid shear stresses than if the flow were unarmoured. This is because each grain, at the moment of impact with the bed, exerts a 'point' force of considerable instantaneous intensity.

1.11 Sediment Transport

Evidently, if a fluid stream can entrain sediment from a bed, then it can also carry along for some distance that entrained material. Thus sediment can be transported by a fluid stream, as we know from common experience. We can also say at this stage that, whatever the mode of sediment transport, the carriage depends ultimately on drag forces exerted by the flowing fluid.

The question of what controls the mode and rate of sediment transport has exercised engineers and scientists for more than a century. Numerous relationships between transport rate and flow quantities have been proposed, many empirical and some theoretical. Unhappily, the empirical relationships are generally found to be inapplicable beyond the limited circumstances of their origin, whilst most of the theoretical ones depend on arbitrary assumptions of little or no physical validity. In recent years, however, R. A. Bagnold has created a body of experimental data and theoretical work showing that sediment transport by a fluid stream may be validly approached from energetics. Namely, we can consider a fluid stream to be a *transporting machine expending power to do work.* For any machine

$$\frac{\text{rate of doing}}{\text{work}} = \frac{\text{available}}{\text{power}} - \frac{\text{unutilized}}{\text{power}}, \tag{1.41}$$

or, in an equivalent alternative form,

$$\frac{\text{rate of doing}}{\text{work}} = \text{efficiency} \times \frac{\text{available}}{\text{power}}. \tag{1.42}$$

We know by observation that the solids ordinarily borne along by fluid streams are more dense than the transporting fluids. Such solids, whether they are momentarily stationary on the bed of the flow or being dragged along by the stream, must therefore be continually affected by a *downward-acting body force* which serves to keep the particles on the bed or pull them towards it. Yet we also know from experience that a steady fluid stream is capable of maintaining a steady transport of sediment, and the question immediately arises as to why the transported particles do not all fall to the bed and become incorporated in it. We can only suppose that the continuously acting downward body force exerted on the moving sediment is in equilibrium with some continuously acting *force directed perpendicularly upward* from the bed of the fluid flow.

Hence each unit area of the bed of a fluid stream carrying sediment is acted on by four forces, two being tangential and two normal (Fig 1.16). The tangential forces are the boundary shear stress τ_0 exerted by the stream of mixed fluid and sediment particles, and the

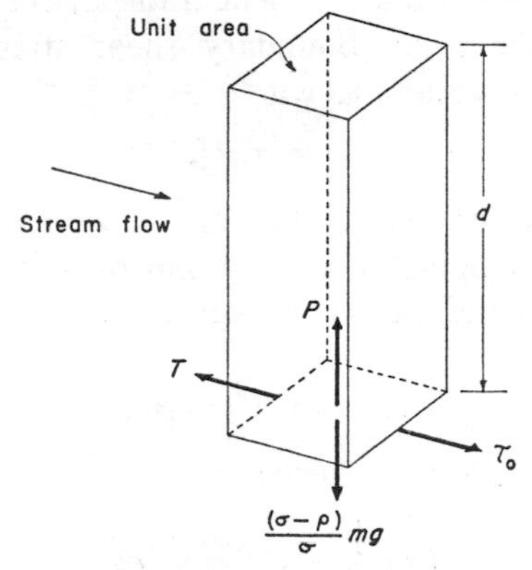

Fig 1.16 Equilibrium of the stream load.

stress T exerted by the stationary bed on the stream. Putting σ and ρ as the solid and fluid densities, respectively, the normal or pressure forces are the load $mg(\sigma-\rho)/\sigma$ acting downwards, where m is the dry mass of the solids above unit area of the bed, and P acting upwards. For equilibrium, $\tau_0 = T$ and $mg(\sigma-\rho)/\sigma = P$. Now when $mg(\sigma-\rho)/\sigma$ is multiplied by the mean velocity at which the particles in the sediment load are transported, we obtain the *sediment transport rate*, denoted by the symbol i, in terms of immersed weight passed per unit time per unit width of bed. But $mg(\sigma-\rho)/\sigma$ has the quality and dimensions of a force and, when combined with the mean sediment load transport velocity, gives us a quantity with the quality and dimensions of a rate of doing work, i.e. of force (per unit area) times distance divided by time. This quantity is, as we have seen, the transport rate i. The transport rate i becomes an actual work rate when multiplied by a conversion factor which is the ratio of the shear stress necessary to maintain the transport to the normal force due to the immersed weight of the load. In summary,

$$\text{sediment load transport rate} = i = \frac{(\sigma-\rho)}{\sigma} mgU_s, \qquad (1.43)$$

and

$$\text{sediment load work rate} = iA = \frac{(\sigma-\rho)}{\sigma} mgU_s A, \qquad (1.44)$$

in which U_s is the mean transport velocity of the sediment load, and A is the conversion factor.

To complete Eq. (1.42) we must define the available power of the fluid stream engaged in the sediment transport. This is found to be the product of the mean boundary shear stress and the mean velocity of the fluid stream, to wit

$$\omega = \tau_0 U, \qquad (1.45)$$

wherein ω is the power and U is the mean flow velocity. We can now equate the power supplied to the column of fluid over unit area of the stream bed to the sediment load work rate. Then Eq. (1.42) becomes

$$iA = e\tau_0 U = e\omega, \qquad (1.46)$$

in which e is the efficiency, whence

$$i = \frac{1}{A} e\tau_0 U = \frac{1}{A} e\omega. \qquad (1.47)$$

This last equation is general, and to use it we have only to infer the efficiency and the flow power.

The conversion factor A entered Eq. (1.44) so that the sediment load could be related to the boundary shear stress responsible for the transport. We should now consider briefly the mechanisms whereby the total force acting on the transported particles might acquire a normal as well as a tangential component.

One mechanism involves the *repeated frequent collision* between grains in shearing motion that extends downward to the unmoved grains on the flow bed. Its nature may be grasped by watching the behaviour of a dozen or so ping-pong balls as they shear over each other and across the flat bottom of a large box slowly shaken from side to side. The balls will be seen to rise up distances of several to many diameters above the bottom of the box, simply as the result of colliding with (and bouncing off) each other and the bottom. Hence in the case of transported sediment the upward acting component of the total force is transmitted directly from the stationary grains on the bed to the grains in shearing motion. The tangential component is likewise transmitted directly from the bed by way of the colliding grains to the fluid stream. Putting T for the tangential component and P for the normal component, experiments with assemblages of densely arranged grains undergoing shearing have shown that T/P ranges between 0·37 and 0·75 according to the conditions of shearing. A region of 'viscous shearing' can be distinguished from a region of 'inertial shearing' by means of a number corresponding to a Reynolds number. In the viscous region, the viscosity of the intergranular fluid determines the collision conditions, the grains as much 'pushing' as 'knocking' each other. In the inertial region, however, the grain mass controls the collision conditions. The grains apparently give each other definite blows, much as the ping-pong balls in the experiment.

The mechanism just outlined requires that the grains be closely arranged and that the normal force component be directly transmitted from the bed. It is therefore applicable only to the coarser grained portion of the sediment load, carried in dense array in close proximity to the bed of the fluid stream. It governs the transport of sand and gravel as *bed load.*

The upward acting normal component of the total force could also be transmitted *indirectly* from the bed by way of parcels of fluid that were moving upwards. This mechanism would be applicable when the sediment particles were so widely spaced apart that they seldom if ever collided. It demands, however, that the fluid stream be turbulent, for there would not otherwise be upward moving fluid elements. The mechanism also demands that, when the momentum of the downward moving elements is compared to that of the upward moving ones, there should be a residual upward momentum flux. For without this we would be in the position of maintaining a steady sediment transport rate without doing work, and perpetual motion would be possible!

This second mechanism is responsible for the transport of the fine sediment in suspension, as a *suspended load* distributed throughout the full thickness of the fluid concerned. As can be seen from Fig 1.15, transport in suspension is the dominant mode of carriage of quartz-density solids whose diameter is less than about 0·015 cm.

Now that some general principles of sediment transport in moving fluids have been established, it is appropriate to inquire a little more closely into the conditions of validity and wider significance of the general transport relation expressed as Eq. (1.47). This equation is, in the strictest sense, valid only in cases of steady uniform flow. Now in steady uniform flow, the fluid velocity at a point remains constant with time and the fluid velocity remains unchanged with distance along a streamline. The geologist may therefore well ask what relevance has Eq. (1.47), strictly valid for steady uniform conditions, to the problem of sediment deposition and erosion during the course of fluid transport. For erosion and deposition evidently express the *varying ability* of fluid flows to bear along sedimentary particles. The clue is, however, to be found in the equation under discussion, because the power available to the fluid stream can be expressed, from Eq. (1.45) and a slightly rearranged form of Eq. (1.26), in terms of the mean fluid flow velocity U. The flow power, and hence the sediment transport rate, are seen to be proportional to U^3. If then the flow is unsteady or non-uniform, or both, then U must vary and the sediment transport rate i must correspondingly change.

These arguments can without difficulty be put in a simple mathematical form. If we take the most general case of all, when the sediment-bearing fluid flow is simultaneously unsteady and non-uniform, then provided the flow does not change too rapidly, the rate of erosion or deposition can be written

$$\begin{array}{l}\text{rate of erosion}\\ \text{or deposition}\end{array} = \frac{\partial i}{\partial x} + \frac{1}{U}\frac{\partial i}{\partial t}, \qquad (1.48)$$

in which the rate of erosion or deposition has the dimensional formula $[ML^{-2}T^{-1}]$, x is distance in the direction of flow, and t is time. The first term of the sum to the right of the equals sign represents the contribution from the non-uniformity of the flow, and the second term the contribution from the unsteadiness. If the flow is steady and uniform, however, each term of the sum is zero, and the sum is therefore itself zero. Whether we get deposition or erosion depends, evidently, on the sign of the sum to the right of the equals sign. If U at an instant of time decreases in the flow direction, then i also diminishes in this direction, and $\partial i/\partial x$ is negative signifying deposition on the bed of the flow. A positive value of $\partial i/\partial x$, on the other hand, means erosion of the bed and implies that U is increasing in the flow direction. Similarly, when the term $(1/U)\ \partial i/\partial t$ is negative, we have deposition and the situation that U as observed at a point decreases with time. Were U measured at a point to increase with time, however, we should have $(1/U)\ \partial i/\partial t$ positive and hence erosion of sediment.

It should be noted that the two terms of the sum in Eq. (1.48) are not bound to have the same sign. Consider, for example, an unsteady flow in a channel whose cross-sectional area increases with increasing distance in the flow direction, making the flow non-uniform also. In this case the first term of the sum in Eq. (1.48) is *always* negative, whereas the second term varies in sign accordingly as the flow in the channel is speeding up or slowing down. Even if the flow is speeding up, a state of deposition can persist in the channel we are discussing, provided both $(1/U)$ and $\partial i/\partial t$ are small enough.

1.12 Some Special Features of Loose Granular Solids

A few moments' play with sugar crystals, a spoon, and a bowl will soon convince one that cohesionless granular solids have certain bulk properties allying them with liquids. It will be found that the sugar crystals can be poured, as can liquids, and that, in a manner resembling liquids, they tend in bulk to the shape of the containing vessel. It will be observed, however, that the free surface of the heaped crystals is sloping, and not horizontal, as would have been the case if a true liquid were in the bowl. But it will also be seen that the crystals cannot be heaped up with the aid of the spoon beyond a certain critical slope angle. If this slope angle is exceeded, an avalanche of crystals flows away, and a new, lower slope is assumed by the surface of the mass. These curious modes of behaviour all depend on the static properties of heaped loose grains and on the dynamics of flowing grain layers.

In thinking about these modes of behaviour in terms of natural materials, such as sand and gravel, it is useful to have in mind some 'ideal' sedimentary particle that can serve as the basis of an analytical model. The sphere has obvious geometrical attractions as the ideal particle, though when considering real materials, it would be truer to nature to choose an ellipsoid, for example, the prolate spheroid.

A major bulk property of granular solids is *concentration*, which may be measured as the proportion of space occupied by the solids, or as the spacing of the solids. We can define the fractional volume concentration as

$$C = \frac{\text{space occupied by solids}}{\text{total space}}, \tag{1.49}$$

where C is the concentration, and the concentration in terms of spacing as

$$\lambda = \frac{\text{mean diameter of solids}}{\text{mean free separation distance between solids}}, \tag{1.50}$$

in which λ is called the 'linear concentration'. The linear concentration is related to the volume concentration by the equation

$$\lambda = \frac{1}{(C_*/C^{\frac{1}{3}})-1}, \tag{1.51}$$

in which C_* is the maximum possible static volume concentration.

Experience teaches that the volume concentration of granular solids in bulk is a variable quantity. If eight equal spheres are arranged at the corners of a cube so as to be in contact (Fig 1.17a), we find that $C = \frac{4}{3}\pi/4\sqrt{4} \approx 0{\cdot}52$. This is the loosest possible regular packing of equal spheres, called *cubical packing*. However, we can take the same spheres and arrange them according to triangles in 'cannon-ball' or *rhombohedral packing* (Fig 1.17b), when it is found that $C = \frac{4}{3}\pi/4\sqrt{2} \approx 0{\cdot}74$. This type of piling is the closest possible regular packing of equal spheres. In the case of equal spheres randomly heaped together, like the particles in a natural sand or gravel, we find that the possible volume concentration varies within the quite narrow range $0{\cdot}64 \geqslant C \geqslant 0{\cdot}60$. We can nonetheless loosely or tightly fill a jar with granular material according to whether we leave the jar undisturbed or tap it during filling, as every housewife knows.

Returning to the experiment with the sugar, it was noticed that the grains could not be heaped up beyond a certain critical slope angle without an avalanche taking place. This angle, denoted by ϕ_i, is the *angle of initial yield* of the material. It is a variable, depending on the character of the granular material, the medium in which sliding takes place, and the nature of the stacking-up process. The angle of initial yield for natural granular materials is found to lie in the approximate range $25° \leqslant \phi_i \leqslant 50°$, but for most materials is 30°–35°. After an avalanche has taken place on a heap of loose grains, the slope assumed by the heap is that of the *residual angle after shearing*, denoted by ϕ_r. Experiment has shown that $\phi_i > \phi_r$ and that ϕ_r is very nearly a constant for any given material. The quantity $\Delta\phi = (\phi_i - \phi_r)$ can be as large as 15°.

Intuitively, one expects the angle of initial yield of granular materials to increase as the grains become more closely packed, i.e. as their concentration increases. A model relating the angle of initial yield and the concentration of random assemblages of equal spheres can be made by supposing that the properties of such an assemblage are statistically represented by a quantity of spheres cubically packed and another quantity rhombohedrally packed, arranged in a proportion to yield the observed concentration. From simple considerations of the geometrical stability in the gravity field of different regular arrangements of equal spheres, we can therefore write

$$\tan \phi_i = k_4 z + k_5(1-z) + B, \qquad (1.52)$$

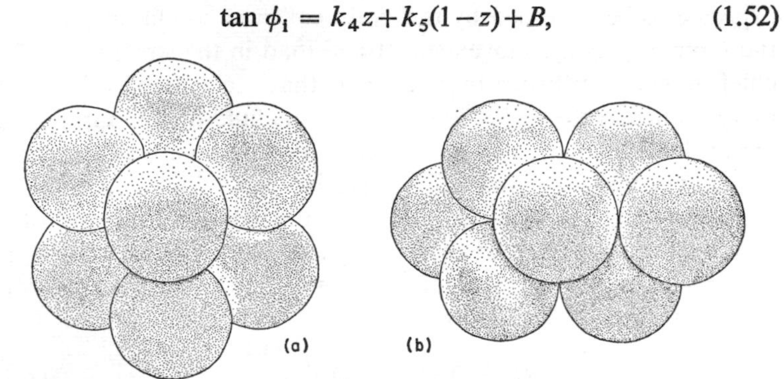

Fig 1.17 Alternative regular sphere packings. (a) Cubical. (b) Rhombohedral.

in which k_4 is the mean tangent of the angle of initial yield of spheres rhombohedrally packed, k_5 is the mean tangent of the angle of initial yield of spheres cubically packed, and B is a necessary dimensionless coefficient representing the sum of forces (e.g. frictional, electrostatic) which might increase or depress the value of $\tan \phi_i$. The quantity z is given by

$$z = \frac{C - C_{cbc}}{C_{rbc} - C_{cbc}}, \tag{1.53}$$

in which C is the observed concentration, C_{cbc} is the concentration of spheres cubically packed ($\approx 0{\cdot}52$), and C_{rbc} is the concentration of spheres rhombohedrally arranged ($\approx 0{\cdot}74$). The quantity z is therefore simply the notional fraction of spheres that are rhombohedrally packed. In Eq. (1.52), which is in good agreement with observation, it has been found that $k_4 = 0{\cdot}7803$ and that $k_5 = 0$. The quantity B is a variable and depends on the surface texture of the grains and the material of which they are made. For example, in experiments with steel ball bearings, it was found that ϕ_i was equal to 25° for the polished balls but 34° for the same balls in a rusty condition. The change in the value of B, then, was about 0·21.

If the slope of an inclined surface formed of granular material is caused to exceed the angle of initial yield, then an avalanche of grains will flow away down the slope, and the angle of the slope will decline to the residual angle after shearing. The grains which constitute the flowing avalanche are certainly highly concentrated, as inspection of an experimental avalanche of sugar crystals will show, but must remain just far enough apart freely to shear over each other. The behaviour of the avalanching mass may therefore be expected to be governed by the grain collision conditions, in much the same way as the movement of bed load in the stream case. The chief difference between these cases is that the avalanche is gravity driven, whereas the bed load of a stream is fluid driven. Grains that are avalanching may, like the material making up a bed load, shear either viscously or inertially. Inertial behaviour is, however, characteristic of mineral-density sand and gravel in air and of all materials coarser than fine sand in water. Very fine and fine sands would shear viscously in water. Putting U_a as the surface speed of a steadily flowing avalanche, we can write for inertial conditions

$$U_a = 0{\cdot}11(g \sin \phi_r)^{\frac{1}{2}} \frac{Y^{3/2}}{D}, \tag{1.54}$$

and for viscous conditions

$$U_a = 0{\cdot}00063\,\frac{(\sigma-\rho)g\sin\phi_r}{\mu}\,Y^2, \qquad (1.55)$$

in which D is the mean diameter of the avalanching grains, Y is the normal thickness of the avalanche, σ and ρ are the densities of the grains and medium, respectively, and μ is the viscosity of the medium.

1.13 A Note on Hydrodynamic Stability

Reference was made above to the 'instability' of certain fluid motions, without further explanation. It is now appropriate to look briefly at this complicated and difficult, yet fundamental, topic in fluid mechanics. The subject is fundamental, because very many natural fluid flows are unstable, but difficult and complicated because of the analytical and physical problems presented. The difficulties are compounded when we consider a flow composed of a fluid mixed with sedimentary particles, in which some forces are due to the fluid phase alone and others to the contained solid phase.

In the real world, all mechanical systems, whether static or dynamic, are subject to small disturbing forces which act to alter the system from some initial state. We can describe a mechanical system as *stable* when the disturbing forces fail to bring about such an alteration, but as *unstable* when the disturbing forces are able to effect the change. The small disturbing forces we have spoken of can be regarded as *destabilizing forces.* Most mechanical systems are, however, acted on by other forces whose role is *stabilizing*. In the case of each mechanical system, then, we can compare the destabilizing to the stabilizing forces, and might expect instability to set in when these forces were in a certain critical ratio. Hence the ratio of the destabilizing to the stabilizing forces is a parameter which will serve as a criterion of stability, and whether any particular system will exist in the stable or the unstable state will depend on the value of the appropriate criterion. These are the basic ideas behind the topic of hydrodynamic stability, in which we consider the stability of mechanical systems composed of fluids, with or without an additional solid phase, and either stationary or in motion. The main objects of an investigation into the stability of a hydromechanical system are the conditions for neutral stability and the physical mechanism of instability.

A very simple case is when a layer of one fluid overlies a layer of another fluid, both fluids being at rest in the gravity field. If the fluid in the upper layer is less dense than that in the lower layer, the stratification is stable against gravitational forces. When, however, the upper fluid is more dense than the lower, the stratification is unstable to gravitational forces and the upper fluid moves down to displace the lower. Gravity provides the mechanism of instability in this case and, provided the upper fluid is the more dense and viscous and surface forces are neglected, instability will always occur.

A somewhat more difficult case is a layer of stationary fluid heated from below or cooled from above. The effect of the heating or cooling is to induce a gravitationally unstable density gradient in the layer of fluid. The fluid is viscous, however, and the viscous forces tend to damp the convective motion called into being by the density gradient. The stability criterion in this instance is the ratio of the destabilizing gravitational force to the stabilizing viscous force, known as the Rayleigh number, where

$$\text{Rayleigh number} = -\frac{abgy^4}{\nu k_6}, \tag{1.56}$$

in which a is the coefficient of thermal expansion of the fluid, b is the temperature gradient, g is the acceleration due to gravity, y is the thickness of the fluid layer, k_6 is the thermal diffusivity of the fluid, and ν its kinematic viscosity. The negative term in Eq. (1.56) is the temperature gradient b, measured upward from the bottom of the fluid layer. If b were not negative, the system would necessarily be stable. The critical value of the Rayleigh number is of the order of $1\text{–}2 \times 10^3$, according to the nature of the boundaries of the fluid layer. When the Rayleigh number is larger than this range, the fluid layer is unstable and convective motions set in. The polygonal cells in which these convective currents appear are known as Bénard cells.

But instability is not confined to hydromechanical systems composed of stationary fluids. A very simple illustration of the instability of fluid in motion is afforded by a slow stream of water flowing from a tap. Close to the tap the stream is a smooth-walled cylindrical column, but at a larger distance is a string of droplets. The instability in this instance is due to the action of surface tension forces on the walls of the emerging cylinder of fluid. These forces cause the cylinder to become pinched laterally at regularly spaced points, with the result that fairly uniformly sized water drops are eventually formed.

A much more important case of the stability of a fluid motion is that of a laminar flow in a channel or past a flat wall. The destabilizing forces in this case are inertial ones, with the viscous forces playing a complicated role. Over a part of the possible range of flow conditions, the viscous forces have a damping role, but over other parts of the range are actually the cause of instability. The criterion for instability of this type of flow is the Reynolds number, which we have already met (see Eq. 1.17) as the ratio of inertial to viscous forces. As we saw, when the Reynolds number for flow is smaller than a certain critical value, the motion is laminar and, we can now say, stable. But the flow is turbulent and unstable when the Reynolds number exceeds the critical value.

The laminar-turbulent transition is a problem in hydrodynamic stability which, because of its great interest and importance, has been the subject of much theoretical and experimental work. The investigations have been mainly concerned with the effect of small disturbances of different physical wavelengths λ on the stability of a laminar flow. The theoretical curve of Fig 1.18, comparing well with experimental results, was obtained for the stability of a laminar boundary layer on a flat plate. On the ordinate of this graph is plotted the group $2\pi\delta^*/\lambda$, in which λ is the physical wavelength of the disturbance and δ^* is a measure of the boundary layer thickness, equal to about one-third the thickness specified by Eq. (1.19). On the abscissa we have the Reynolds number based on the boundary layer thickness δ^* and the free stream velocity U_∞. The curve shown in the figure is the so-called 'neutral curve' dividing the stable laminar from the unstable flow. It will be seen that the minimum Reynolds number for unstable flow is about $Re = 420$. It will also be noticed that for small enough values of λ the flow is laminar at all Reynolds numbers.

There are many instances in nature where a fluid is obliged to travel over a curved path, for example, a stream in a meandering channel. The flow of a fluid over a curved path is unstable to centrifugal forces if the property of the fluid known as the circulation decreases outward from the centre of curvature of the motion or if the pressure gradient does not balance the centrifugal force. The circulation is simply the product of the velocity of a fluid element and

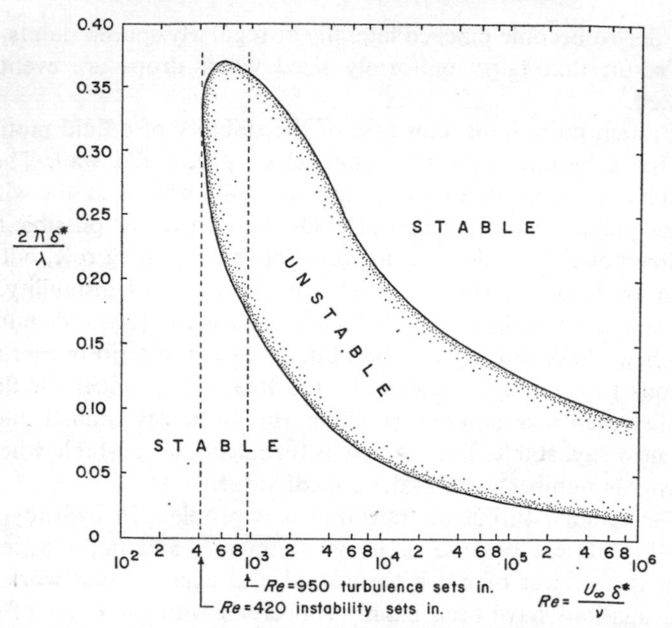

Fig 1.18 Curve of neutral stability of the boundary layer on a flat plate at zero incidence in a uniform stream (after W. Tollmien). Note that turbulence sets in at a lowest critical Reynolds number somewhat more than twice as large as the lowest critical Reynolds number for instability. Instability is in this case marked by the appearance in the flow of various wave-like features.

the radius of curvature of its path of travel. For example, when we stir a beaker of water containing a little sand, we find that the sand is dragged to the centre of the bottom of the beaker. This is because the water in motion is unstable to centrifugal forces, so that each element, instead of taking a circular course round the beaker, travels over a helical spiral path directed so that, at the bottom of the beaker, the fluid moves inward toward the centre. We have, in effect, created an infinite stream meander. A similar unstable motion can be generated in the flow over a concave wall, when spiral vortices with axes parallel to flow are set up in the flow close to the wall. Vortices of this particular type we have called secondary flows, but they will also be met with under the name of Taylor-Görtler vortices. They appear to influence the character of several different sedimentary structures.

When a fluid flow carries sediment along with it, we have to consider not only the forces associated with the motion of the fluid, but also those arising from the presence of the sediment. We know very well that many natural two-phase flows are unstable, but have thus far been able to make little progress with them owing to their great complexity. One point is clear, however. The quantities which govern a two-phase flow are sufficiently numerous that several distinct unstable modes of flow can simultaneously exist in any one system. River meanders, underwater dunes, and current ripples manifest distinct states of unstable flow of a two-phase fluid. Yet in any one river we can find current ripples superimposed on underwater dunes, and underwater dunes superimposed on meandering channels, all in simultaneous existence and movement.

READINGS FOR CHAPTER 1

This chapter, in a small number of pages, covers in a simple way topics that are otherwise dealt with in many large books and a number of already distinct disciplines. A list of selected papers is out of the question, no matter how ruthlessly it was prepared, and the best that can be done is to list a few works of help in strengthening physical intuition or useful for reference.

An elegant and elementary introduction to some of the basic concepts of fluid mechanics is given by:

SHAPIRO, A. H. 1964. *Shape and Flow.* Heinemann, London, 186 p.

though his definition of viscosity is misprinted. Useful for reference are:

KAUFMANN, W. 1963. *Fluid Mechanics.* McGraw-Hill Book Co., New York, 432 p.

PRANDTL, L. 1952. *Essentials of Fluid Dynamics.* Blackie and Son, London, 452 p.

RAUDKIVI, A. J. 1967. *Loose Boundary Hydraulics.* Pergamon, Oxford, 331 p.

SCHLICHTING, H. 1960. *Boundary Layer Theory*, 4th ed., McGraw-Hill Book Co., New York, 647 p.

A fundamental reference on sediment transport is:

BAGNOLD, R. A. 1966. 'An approach to the sediment transport problem from general physics.' *Prof. Pap. U.S. geol. Surv.*, 422-I, 37 p.

CHAPTER 2

Some Sedimentary Structures, and a Note on Texture and Fabric

2.1 General

Clastic sediments abound in a great variety of structures, some due to currents, others to post-depositional movement of one bed relative to another, and still others to organisms. There are certain of these structures of disproportionately large physical interest compared with their value in geological interpretation, and others that are environmentally diagnostic but physically trivial. Happily, physical interest very often coincides with interpretative importance. In this chapter we shall briefly consider a selection of structures which combine these features in an illuminating way, though many more structures will be encountered in the book as a whole. Attention will also be given to the origin of sediment sorting and fabric.

2.2 Classification of Structures

Their very number and variety makes sedimentary structures difficult to classify, though three principal classes can be recognized (Appendix II). *Biogenetic* structures, such as tracks and tunnels, are due to the activities of organisms and of no further concern here. Those called *endogenetic* result from the action of mechanical forces which arise within sediments after deposition but before lithification. These structures record the movement of one part of a deposit relative to another, and include load casts and convolute laminations. *Exogenetic* structures are due to the operation, at the sediment-fluid interface, of external forces. They comprise the largest and most diversified group of structures, and demand closer attention.

Three groups of exogenetic structures are recognized (Appendix II). *Surface markings* are due to forces unconnected with currents acting tangentially on the bed. This group is a mixed bag, and includes suncracks as well as sand volcanoes. *Bed forms* are spatially periodic mounds and hollows fashioned at the sediment-fluid interface by the action of tangential fluid forces either unipolar, bipolar or multipolar in direction of action. The bed material may be either cohesionless sand or gravel, or cohesive mud. When bed forms arise on a cohesive bed, material is ordinarily transferred only from bed to flow, i.e. there is erosion only. Transfer in the cohesionless case is always from flow to bed as well as from bed to flow. It is because deposition—or transfer from flow to bed—occurs locally from bed forms travelling over a cohesionless bed, that *internal structures* such as cross-stratification are produced. Most of the internal structures distinguished in Appendix II are due to bed-form migration. Bed forms can also be classified according to their maximum elongation relative to the direction of the parent current. Forms elongated *transversely* to flow are easily seen to be adjusted to a two-dimensional unstable flow in which flow quantities are perturbed *parallel* to flow. The forms arranged *parallel* to flow are a response to a three-dimensional or secondary flow, in which flow quantities are perturbed *transversely* to flow.

2.3 Parting Lineations

When finely crushed coal lightly sprinkled into a sink is gently flushed out, the particles become grouped into long ridges parallel to flow between which are clear lanes. This structure closely resembles the parting lineations seen as *en echelon* ridges and hollows on the bedding of evenly laminated sandstones. The ridges are 10–15 cm long but scarcely a few sand grains in height, while the hollows are flat bottomed and several millimetres to a centimetre or two wide. The grains in the ridges are somewhat coarser than beneath the hollows, and lie with their long axes parallel on the average to the trend of the macroscopic lineation. The long axes are not distributed unimodally, however, for the two strongest modes, spaced 30°–40° of arc apart, lie symmetrically about the value of the statistical average orientation.

Parting lineations were once thought to be restricted to a mode of aqueous sediment transport typified by a plane bed of wide extent, such as occur in the beach swash-zone and on certain river bars. Later work showed that they arise during transport across any surface substantially plane on a scale larger than the lineations, provided always that the sand is not too coarse and there is bedload movement. The lineations appear on the backs of active sand ripples and dunes, where there is erosion, and on antidune beds as well.

The experiment with coal is illuminating about parting lineations. Since we began with grains randomly dispersed, and ended with them in ridges, there must have existed close to the bed transverse components of flow which, in the course of the general transport, dragged the grains from the clear lanes towards the ridges. The skin-friction lines therefore formed a symmetrical herringbone pattern (Fig 2.1), reminiscent of a secondary flow. Now the flow structure made evident by the grains is experimentally a universal feature of the laminar sublayer and lower buffer layer in the turbulent boundary layer, and could equally well have been revealed using dye. The mean transverse distance apart of the grain-ridges can therefore be regarded as identical with the mean transverse scale of the flow structure, given empirically by

$$\lambda_z = \frac{100\nu}{U_*}, \tag{2.1}$$

in which ν is the kinematic viscosity and U_* the shear velocity. The values of λ_z given in Fig. 2.1 for an aqueous flow over a hydraulically smooth boundary cover the same range as the spacing of lineations observed from the field. Since λ_z decreases with increasing U_* we can begin to see why parting lineations are uncommon in medium and coarser sands. These sands are generally moved by relatively strong rough flows, when the calculated scale of the lineations is of the same order as the size of the largest grains in transport.

2.4 Sand Ribbons

Sand meagrely supplied to an inerodible substrate is ordinarily transported in the form of long, *en echelon* ribbons arranged parallel to flow at a roughly constant transverse spacing. In the desert, the ribbons are many metres apart across the wind and hundreds of metres long. They can be either smooth or covered with wind ripples, though their relief never exceeds a few centimetres. Sand ribbons can form on the gravel beds of shallow rivers and, where currents are strong, on the rocky or stony floor of the sea in depths up to 100 metres. Transverse current ripples frequently appear on the smaller ribbons, but on the larger ones low dunes are sometimes found. The transverse spacing of ribbons formed in water varies from one to four times the flow depth. Their length is generally 10 to 20 times the transverse spacing.

The orientation of sand ribbons betrays them as the response of transported sediment to a mode of secondary flow. Arguing as for parting lineations, which the ribbons clearly resemble, we would not expect strips of sand with clear lanes between unless transverse as well as downstream components of flow existed close to the bed. Indeed, the subordinate structures often preserved on the ribbons allow skin-friction lines to be plotted in a symmetrical herringbone pattern similar to that mapped for parting lineations (Fig 2.1). Continuity then demands that there exists in the flow a secondary motion in the form of oppositely rotating cellular vortices with axes parallel to flow. The vortices appear to be square or rectangular in transverse section and probably occupy the full flow depth. The instability giving rise to the vortices probably depends on inequalities

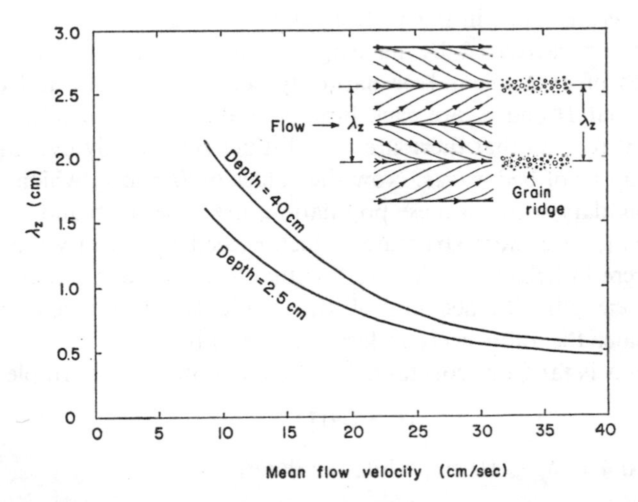

Fig 2.1 The mean transverse spacing of parting lineations and kinematic structural features of the near-bed flow as a function of mean flow velocity and flow depth. A hydraulically smooth boundary is assumed.

between the turbulent shear stress components, and may not depend on the character of the bed.

2.5 Ripples and Dunes in Water

Ripples and dunes are plentiful wherever comparatively gentle flows of water drive along loose sand. Equally common in sedimentary deposits are internal structures which resulted from the travel of ripples and dunes, namely, types of cross-stratification characterized by a succession of units of a similar thickness and foreset orientation. Ripples occur on river beds, on beaches, and on the floor of the deep sea. Dunes abound in large rivers, in the channels of tidal estuaries, on tidal sand banks, and on the sandy floors of shallow, tide-swept seas.

Ripples and dunes are transverse, wave-like structures which occur in trains of similar individuals wherever flow conditions are sufficiently uniform. Both types of structure are asymmetrical in vertical profile parallel to flow. The leeside, facing downcurrent, is swept by avalanches which originate at the crest and slopes down at about 30° from the horizontal. The stoss-side, facing the flow, assumes a slope between about 1° and 8° depending on the size of the structure. Thus in the vertical plane parallel to flow a ripple or dune is characterized by its height H and wavelength λ_x. Large samples of experimental or naturally occurring ripples and dunes reveal that H and λ_x are each bimodally distributed in value, from which it follows that there are two distinct but slightly overlapping populations of bed waves. Now the values of H and λ_x which typify the boundary between these populations are $H \approx 4$ cm and $\lambda_x \approx 60$ cm. Dunes are those structures which exceed 60 cm in wavelength and 4 cm in height; ripples are less than 4 cm in height and 60 cm in wavelength. On account of the gentle slope assumed by the stoss-side, the ratio λ_x/H is large, commonly between 10 and 20, though it is far from constant. Field data show that for ripples

$$H = 0{\cdot}074\lambda_x^{1{\cdot}19}, \tag{2.2}$$

where $0{\cdot}4 \leqslant \lambda_x \leqslant 60$ cm, while for dunes

$$H = 0{\cdot}074\lambda_x^{0{\cdot}77}, \tag{2.3}$$

where $0{\cdot}60 < \lambda_x \leqslant 2000$ metres. Thus ripples can be very small indeed, whereas the largest aqueous dunes resemble desert dunes in scale.

Although different in size, ripples and dunes are similar in shape in plan as well as profile. The structures are called long-crested if the crest extends across the flow for a distance several to many times the wavelength. Some long-crested ripples and dunes have practically rectilinear crests, though others have sinuous crest lines which turn smoothly from side to side. The crests of long dunes are occasionally shaped like catenary waves, the sharply pointed segments indicating the flow direction. Many ripples and dunes have, however, a breadth of the same order as the wavelength and short, strongly curved crest lines. The commonest dune of this sort is called lunate and has a crest which is concave in the direction of flow. A linguoid ripple has a strongly curved crest line convex in the direction of the current.

Since ripples and dunes travel with the flow, the inclined foreset layers or cross-strata preserved in sequence within their bodies record the successive earlier position of the leeside. Where these layers are relatively thick ($\approx$ 5–15 mm), as in a dune, they are called cross-beds. The inclined layers or cross-laminae found inside ripples are seldom thicker than a millimetre, however.

The extent of fossilization of cross-strata depends partly on the variation of sediment transport rate along the flow. If a ripple or dune of constant height travels parallel to its base, then the quantity of sediment removed in unit time from the stoss-side equals the quantity deposited in the same time on the leeside. We see from Fig 2.2a that

$$\underset{\text{(eroded from stoss-side)}}{\text{area } ABEF} = \underset{\text{(deposited on leeside)}}{\text{area } ACEG} \tag{2.4}$$

The mass rate of transport i_W of sediment involved in the bed-wave movement is easily found to be

$$i_W = \tfrac{1}{2} H U_W \gamma, \tag{2.5}$$

where U_W is the wave velocity relative to the ground and γ is the sediment bulk density. Beneath the wave there occurs neither deposition nor erosion, as measured on a scale larger than the wave, and so the transport rate remains constant at this scale along the line of flow. A cross-stratified deposit would be preserved, together with the form of the bed wave, only if the wave was stopped and rapidly buried. But if the path of travel of a ripple or dune is not

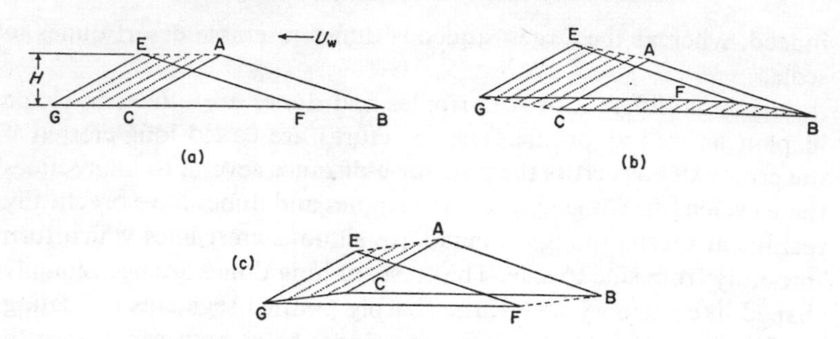

Fig 2.2 Equilibrium of a ripple or dune beneath a unidirectional flow from right to left. (a) Uniform steady motion. (b) Net deposition. (c) Net erosion.

parallel to its base, then a state of either net deposition or net erosion may exist, again as measured on a scale larger than the wave. The wave of Fig 2.2b moves on a path inclined above its base, the height remaining constant, whence

area *BCFG*	=	area *ACEG*	−	area *ABEF*	(2.6)
(net deposit, cross-stratified)		(deposited on leeside)		(eroded from stoss-side)	

But since the height is constant, and area *ACEG* > area *ABEF*, a quantity of sediment proportional to area *BCFG* must have settled on the wave from the flow above during the time of travel. The cross-stratified deposit in area *BCFG* is preserved within the bed even though the wave moves onward, as can be seen from Fig 2.2b. Conditions favouring this occur where a flow increases in depth in the flow direction and thus gradually loses its ability to transport sediment. For both flow velocity and bed shear stress, and hence stream power, decrease with increasing depth for constant discharge and flow width. Favourable conditions also arise where depth remains constant but flow velocity decreases. Now the wave of constant height of Fig 2.2c follows a path inclined below its base, and net erosion occurs overall, since

area *BCFG*	=	area *ABEF*	−	area *ACEG*.	(2.7)
(net erosion)		(erosion from stoss-side)		(deposited on leeside)	

Net erosion will occur where ripples or dunes travel with a flow increasing in power downstream, or which increases in velocity for constant depth at a point.

We can render the situation of Fig 2.2 quantitative by considering what happens when net deposition or net erosion occurs on a flat-lying bed covered with ripples or dunes of uniform height and wavelength. When the bed waves are in motion, they travel along a definite path with a definite velocity. In the general case, let the bed forms travel with a velocity U along a path inclined at an angle ζ relative to the base of the bed form parallel with the generalized bed. The velocity U can be resolved into a component parallel with the generalized bed, and another component normal to the generalized bed. The component parallel with the bed is the velocity U_W introduced in Eq. (2.5). Let the normal component be U_y. In Fig 2.2a, for example, $\zeta = 0°$ and therefore $U_y = 0$. In the other two cases, however, ζ is not zero and therefore neither is U_y, which is measured upward in the case of the climbing ripples of Fig 2.2b, but downward in the case of the descending ripples or dunes of Fig 2.2c.

Now U_y is evidently proportional to the rate at which sediment is transferred between bed and flow across unit area of the generalized bed. That is, when measured upward, U_y records net deposition on the bed, but when measured downward it denotes net erosion over the bed. Denoting this rate of transfer between bed and flow by R, negative when erosion occurs but positive when deposition takes place, we can write

$$U_y = \frac{R}{\gamma}. \tag{2.8}$$

Putting $\tan \zeta$ in terms of U_W and U_y, it is readily found from Eqs. (2.5) and (2.8) that

$$\tan \zeta = \frac{RH}{2i_W}, \tag{2.9}$$

stating that the angle of the path of the bed wave is directly proportional to wave height and the net rate of erosion or deposition, but inversely proportional to the bedload transport rate. A very little consideration will show that the sediment transfer and transport rates differ in dimensions, and therefore differ physically. The transfer rate, i.e. the net rate of erosion or deposition, is in fact in mathematical terms dependent on the partial differential coefficients of the transport rate (see Eq. 1.48).

The above model of cross-stratified deposits can be completed by adding the third dimension from a knowledge of the shapes of ripples and dunes in plan, net deposition being assumed (Fig 2.3). Ripples and dunes with long, straight crests necessarily give rise during travel to very nearly plane cross-strata and to very nearly tabular cross-stratified units. As the crests shorten, and as curved crestal segments appear, the bed waves lay down curved cross-strata which fill trough or scoop-shaped erosional hollows. The transverse dimension of a cross-stratified unit, and the degree of curvature of its base, therefore depend on the overall shape of the bed wave. The extent of net deposition substantially affects the geometry of a cross-stratified deposit only when a bed wave fails completely to behead its predecessor. Ripples often climb up such steeply inclined paths that laminae are preserved on the stoss-sides.

The processes acting to form the cross-strata themselves are rather complicated. The grains continuously discharged over the wave crest are convected and diffused in a turbulent, free shear-layer before finally settling on the leeside, at a rate given empirically by

$$\frac{W_x}{W_a} = \left(\frac{x}{a}\right)^{-n}, \tag{2.10a}$$

in which

$$n \propto \frac{V_0}{U}, \tag{2.10b}$$

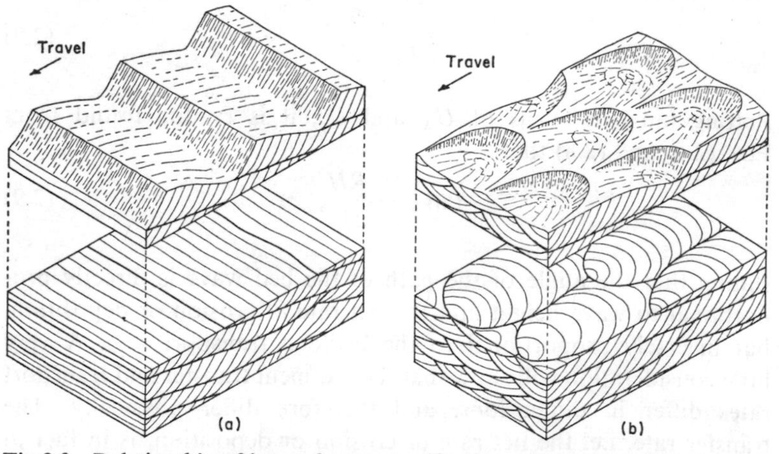

Fig 2.3 Relationship of internal cross-stratification pattern to surface shape of ripples or dunes. (a) Tabular cross-stratification due to bed forms with long and substantially straight crests. (b) Trough cross-stratification due to bed forms with short curved crests, as illustrated by lunate dunes.

where W_x is the settling rate at horizontal distance x downstream from the crest, W_a is the rate at a reference distance a, U is the flow velocity, and V_0 is the sediment free falling velocity. Thus settling increases the slope of the leeside. But an avalanche of grains leaves the crest every time settling buildings up the slope to the angle of initial yield (ϕ_i) of the sediment. The leeside after avalanching is inclined at the smaller, residual angle after shearing (ϕ_r), whence avalanching decreases the slope. Combining these two rates of slope change, we deduce that the observed period of avalanching is

$$P_{obs} = P_\infty + T_0 \left(\frac{P_{obs}}{P_\infty} \right)^m, \tag{2.11a}$$

where

$$P_\infty = \frac{(\phi_i - \phi_r)}{\Omega}, \tag{2.11b}$$

and

$$T_0 = \frac{L}{U_a} \tag{2.11c}$$

in which Ω is an angular velocity proportional to dW/dx, U_a is a characteristic velocity for the avalanches when $\Omega = 0$, L is the distance travelled by the avalanches, and m ($1 > m > -5$) is an empirical exponent. The avalanching is intermittent at small sediment transport rates, but at sufficiently large rates there is a continuous, general sliding of sediment down the lee. The grains carried down by the avalanches become simultaneously sorted as to size, and it is chiefly the sorting which allows one cross-stratum to be distinguished from another. However, the sorting becomes progressively poorer as the period of avalanching falls within the intermittent range, and under conditions of continuous sliding is so poor as scarcely to allow distinction of an internal structure. Inspection of Eq. (2.11) will also show why the thickness of the cross-stratum increases with size of bed wave.

Natural cross-strata assume two principal shapes in profile parallel to flow. The cross-stratum is called angular if it meets the unit base at a sharp, acute angle. The sediment deposited from avalanches then extends to the bottom of the cross-stratum, the whole constituting a foreset layer. When the contact is asymptotic we describe the cross-stratum as tangential. The foreset now extends down only as far as the slope-break at the top of the asymptotic portion of the cross-stratum. The sediment in the bottomset below the slope-break is finer grained than in the foreset deposited from avalanches. An explanation of these features lies in Eq. (2.10) above. For when n is small, practically all grains settle near to the wave crest, so that the avalanches flow unimpeded to the bottom of the lee. As n grows larger, however, the grains settle over increasingly large distances from the crest, to form increasingly long and thick bottomsets. Thus the effect of increasing n is to increase in the cross-stratified unit the proportion of bottomsets relative to foresets. The bottomset deposits are finer grained than the foreset because n is inversely proportional to grain size.

The physical mechanisms governing the size and shape of ripples and dunes are far from being well understood. Taking an aloof view, however, the case is one of instability, for it will be readily apparent that the properties of both the grain and fluid flow vary periodically in the current direction. Dimensional reasoning suggests that λ_x for ripples and dunes is given by

$$\frac{\lambda_x}{D} = f_1\left(\frac{U_* D}{\nu}, \frac{d}{D}, Q\right), \tag{2.12}$$

in which f_1 means 'a function of', D is the grain diameter, d is the depth and flow, and Q is a dimensionless quantity representing the sediment transport rate.

If the flow occurs over a hydraulically smooth boundary, the relative roughness term d/D vanishes and

$$\frac{\lambda_x}{D} = f_2\left(\frac{U_* D}{\nu}, Q\right), \tag{2.13}$$

whence λ_x should be independent of flow depth. Experimental data (Fig 2.4) show that this is approximately true for ripples. It is seen that

$$\lambda_x \approx 1000D, \tag{2.14}$$

but, putting

$$\frac{\lambda_x}{D} = \zeta \frac{d}{D}, \tag{2.15}$$

where ζ is a proportionality constant, we also see that (Fig. 2.5)

$$\zeta \propto \frac{U_* D}{\nu} \quad \left(\frac{U_* D}{\nu} < 10\right). \tag{2.16}$$

The wavelength of ripples therefore depends chiefly on grain size, but is also influenced by flow depth and grain Reynolds number.

The Reynolds number term theoretically becomes irrelevant for hydraulically rough flow, so that

$$\frac{\lambda_x}{D} = f_3\left(\frac{d}{D}, Q\right), \tag{2.17}$$

from which it may be deduced that

$$\lambda_x \approx 5d. \tag{2.18}$$

Fig 2.4 shows that λ_x indeed increases with d for the experimental dunes, though the constant of proportionality is nearer 10 than 5. The range of experimental values of d is small, unfortunately, and field observations suggest that

$$\lambda_x = 1{\cdot}16d^{1{\cdot}55} \tag{2.19}$$

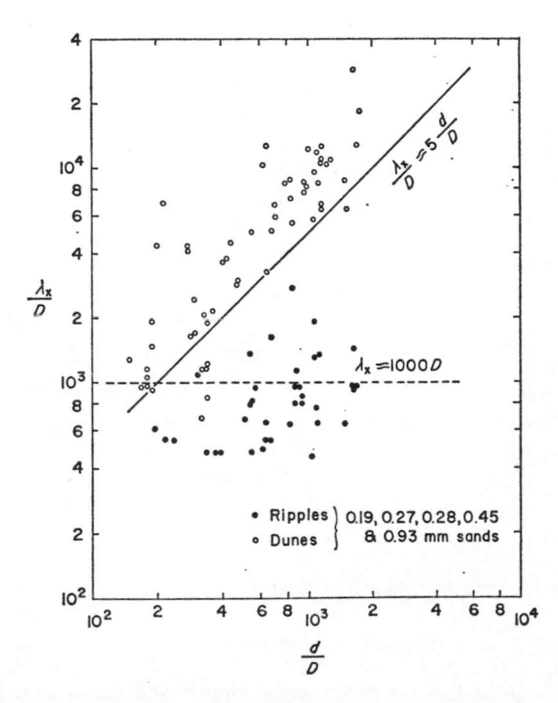

Fig 2.4 Wavelength parallel to flow of experimental ripples and dunes in relation to flow depth (data of H. P. Guy, D. B. Simons and E. V. Richardson).

is a better approximation, where $0{\cdot}1 \leqslant d \leqslant 100$ metres. The same data show that dune height varies with water depth as

$$H = 0{\cdot}086d^{1{\cdot}19}, \tag{2.20}$$

for $0{\cdot}1 \leqslant d \leqslant 100$ metres. Thus the height and wavelength of dunes increase with flow depth, the wavelength growing faster than height (see Eq. 2.3). The wavelength of the experimental dunes (Fig 2.5) is independent of U_*D/ν only when D is small, and there is no clear-cut value of U_*D/ν at which ripples as we have defined them can be distinguished from dunes.

The relationship of ripples and dunes to sediment transport rate is best discussed in terms of stream power, to which the transport rate is proportional. With ascending stream power (Fig 2.6), very fine and fine sands give first ripples, then dunes, and finally a plane bed with parting lineations. Sands whose median fall diameter exceeds about 0·65 mm give first a plane bed, then ripples, and

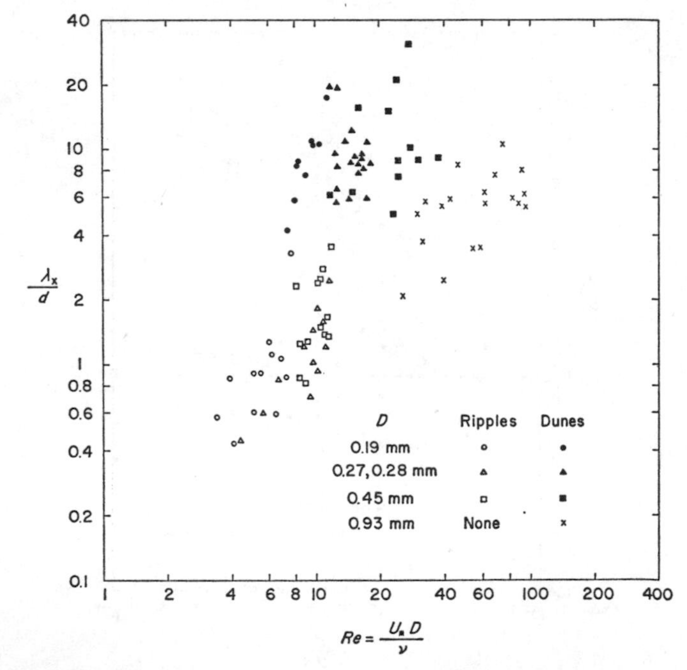

Fig 2.5 The ratio λ_x/d for experimental ripples and dunes as a function of Reynolds number based on shear velocity and sediment size (data of H. P. Guy, D. B. Simons, and E. V. Richardson).

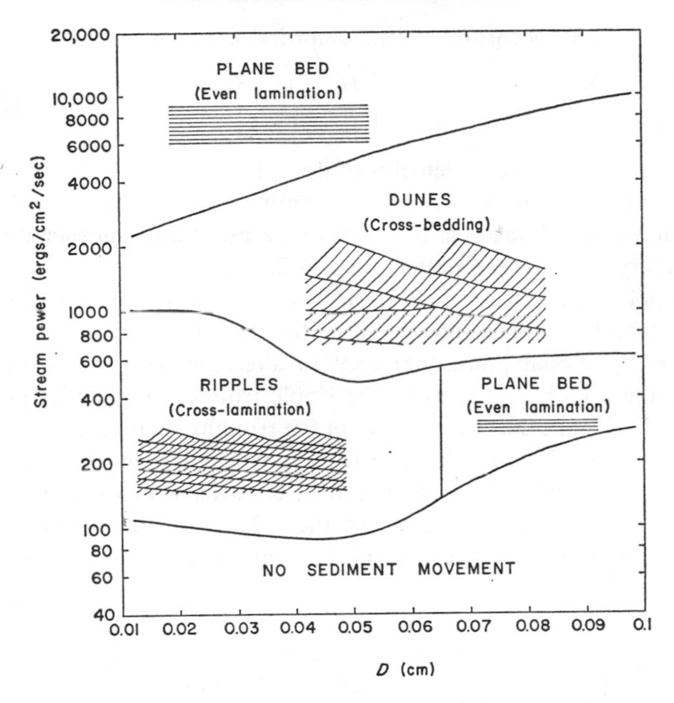

Fig 2.6 Bed form in relation to stream power and calibre of bed-material load (data of G. P. Williams, and H. P. Guy, D. B. Simons and E. V. Richardson).

finally another plane bed over which the transport is intense. The fact that the range of stream power occupied by ripples becomes larger relative to dunes with falling grain size allows us to explain why ripples and cross-lamination are the commonest structures of the finer sands while dunes and cross-bedding are typical of the coarser sand-grade deposits. The same inter-relationship leads to physically meaningful deductions about the physical environment of deposition of fossilized sediments. Moreover, estimates of minimum water depth can be obtained where field evidence allows us to put cross-bedding thickness instead of dune height into Eq. (2.20).

The crests of ripples and dunes shorten and curve more strongly with increasing Froude number of flow and ratio of wave height to water depth. At the same time, ridges and furrows aligned parallel to flow become more closely spaced and conspicuous in the troughs and on the stoss-sides of the bed waves. Thus a transverse dimension λ_z characteristic of the bed waves changes simultaneously with the wave-length λ_x, according to the empirical expression

$$\frac{\lambda_x}{\lambda_z} = 6{\cdot}4\left[\frac{H}{d}\cdot\frac{U}{\sqrt{(gd)}}\right]^{0\cdot27}, \tag{2.21}$$

in which U is the mean velocity of flow. The pattern of skin-friction lines on a ripple or dune bed in fact combines a pattern typical of a two-dimensional unstable flow with separation and another due to secondary flow. In the pattern of Fig 2.7 the crests and troughs of the ripples are expressed as transverse separation and attachment lines respectively. The spacing along the flow of, say, the separation lines is λ_x. Nodal points are seen to alternate with saddle points along the attachment lines in the ripple troughs. The nodal points coincide with the deepest hollows in the troughs, as is easily deduced from the fact that skin-friction lines radiate away in all directions from such points. The saddle points, towards which certain skin-friction lines are directed, lie on the ridges parallel to flow. The ridges occur because the rate of erosion of sediment at a saddle

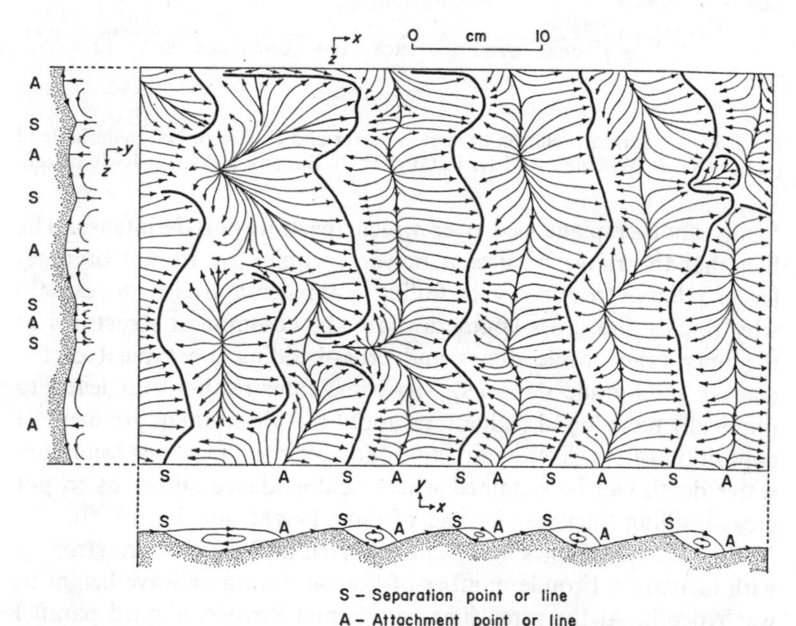

Fig 2.7 Skin-friction lines and streamlines associated with a portion of a bed of experimental ripples in fine-grained quartz sand. Mean flow velocity 22 cm/s, from left to right. Mean flow depth 9·5 cm.

point cannot be as large as at a nodal point, on account of the movement of grains symmetrically towards the point. The spacing across the flow of saddle points is a measure of λ_z.

2.6 Antidunes

The celerity c of simple harmonic gravity waves of length λ on still water of real depth d is given by

$$c^2 = \frac{g\lambda}{2\pi} \tanh \frac{2\pi d}{\lambda}. \tag{2.22}$$

If the water is assigned a flow velocity $U = c$ acting in the opposite direction to c, the wave profile is arrested relative to the ground, and we have

$$U^2 = \frac{g\lambda}{2\pi} \tanh \frac{2\pi d}{\lambda}. \tag{2.23}$$

Now suppose the bed is sand that can be moved by the flow. If the bed is eroded into waves identical in phase and amplitude with the now stationary surface waves, the bottom profile becomes identical with a streamline for the waves as if they existed on an infinitely deep flow. A virtual depth $d_v \to \infty$ now replaces the real depth, whence Eq. (2.23) becomes

$$U^2 = \frac{g\lambda}{2\pi}, \tag{2.24}$$

since $\tanh 2\pi d_v/\lambda \to 1$ as $d_v \to \infty$. Eq. (2.24) states that the wavelength of the bed and surface waves is dependent only on U.

The natural waves whose behaviour corresponds to Eq. (2.24) are called antidunes, because they commonly travel against the current. Antidunes are sinusoidal bed and water waves of low amplitude that are broadly in phase with each other. They arise in free-surface flows that lie in or close to a supercritical condition (i.e. the Froude number is about one), and can be seen in shallow alluvial rivers as well as in gutters after rain and in beach rills. The antidunes of beach rills and gutters seldom exceed a few centimetres or a decimetre or two in wavelength. Those found in rivers, however, reach 5–10 metres in wavelength and up to 2 metres in height. Antidunes cannot exceed a certain critical height without breaking, after which the water surface stays smooth and flat for a brief period. The growth, breaking and reformation of antidunes in a channel reach is commonly rhythmic. Antidunes are ordinarily found where the channel slopes steeply and the sediment load is large.

Field and laboratory studies show that Eq. (2.24) is closely obeyed by antidunes. Since the waves observed in the field are associated with a rather narrow range of Froude numbers distributed about $Fr = 1$, we can also write

$$\lambda \approx 2\pi d \qquad (Fr \approx 1), \tag{2.25}$$

though it should be remembered that λ is not uniquely determined by flow depth.

The internal lamination of antidunes is poorly developed. The laminae are gently inclined, and can lie on either the upstream or downstream side of the bed wave accordingly as the wave migrated upstream or downstream.

2.7 Flute Marks

Flute marks are the commonest scour markings. They are heel-shaped hollows, generally between 0·5 and 2·0 cm in amplitude, abundantly preserved as moulds on the soles of sandstones thought to have resulted chiefly from turbidity current action. The marks lie parallel to flow and in plan vary at the upcurrent end from narrow and sharply pointed to broad and rounded. The longitudinal profile is strongly asymmetrical, with the deepest part of the mark at the upcurrent end.

Natural flutes are due to the erosion of mud beds by sand-laden currents. The marks cut laminae in the mud below and commonly have a discordant fill of laminated sandstone. Sedimentary structures preserved within the overlying bed generally have the same orientation as the marks. Since the mud bed can be supposed initially flat, it follows that each mark grew by differential erosion, in the course of which mud was removed at a greater rate from within the mark than from the surrounding bed. Hence the time T required to develop a mark of amplitude A in a vertically uniform mud of bulk density γ is

$$T = \frac{\gamma A}{(k_1 - 1)E_0}, \tag{2.26}$$

in which E_0 is the rate of erosion of the bed surrounding the mark, and k_1 is the ratio E_i/E_0 where E_i is the rate of erosion at the bottom of the mark. Experimental relationships between erosion rate and boundary shear stress suggests that flute marks of conventional amplitude can be developed in a matter of tens of seconds, or a few minutes, provided the fluid forces acting are large. The value of k_1 is from experiments commonly between 1·2 and 1·5.

Although flutes vary considerably in shape and size, their formation is always associated with processes of flow separation and attachment. Each flute is an abrupt step down in the bed, and so should generate a separated flow, the rim of the mark being the separation line. This has proved true for both narrow and broad flutes (Fig 2.8). The separation bubble present in a broad flute has a closed, proximal portion or roller associated with a nodal attachment point and, distally, an open part or vortex against each flank of the mark. The open parts have corkscrew-like streamlines which pass out of the mark to blend gradually into the external flow. However, the bubbles of narrow flutes are fully open and consist of two oppositely rotating corkscrew vortices. Flutes of asymmetrical plan generate bubbles in which one of the vortices is either relatively small or wholly suppressed.

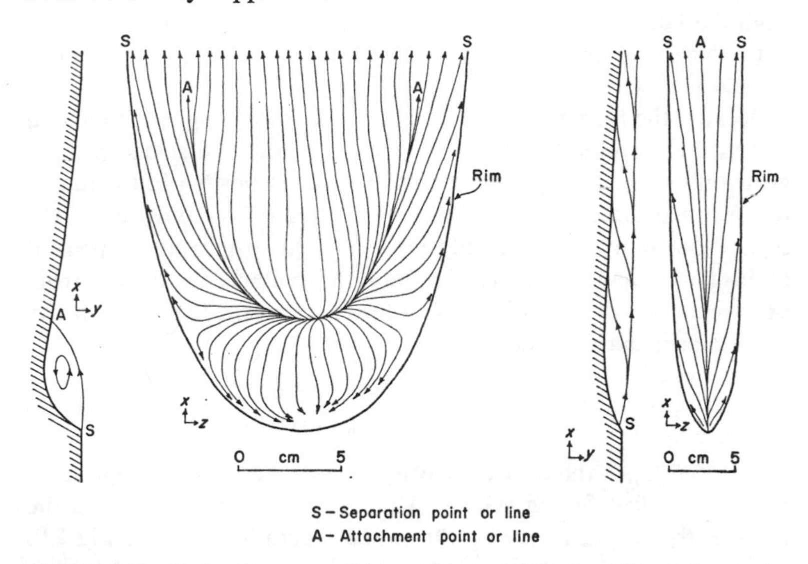

Fig 2.8 Skin-friction lines on experimental flute marks at large Reynolds number based on flute amplitude. External flow from bottom to top.

There follows a physically plausible explanation of flutes in terms of some further properties of separated flows. Since it can reasonably be assumed that the case is one of turbulent flow of large Reynolds number relative to the mark, it follows from experimental results that the free shear layer associated with the separated flow should be of larger turbulence intensity than the flow established on the surrounding bed. As shear stress is proportional to turbulence intensity, and the rate of erosion of mud increases with the applied stress, erosion should proceed faster from within a flute than from the bed round about. The largest rates of erosion would be expected in the region of reattachment, since here eddies generated in the free shear layer first strike the bed.

This hypothesis has been experimentally verified for flute marks and related structures developed on weakly cohesive muds. In this case the stresses exerted by the plain fluid alone are sufficient for the required differential erosion of an initially plane mud bed. The threshold stress for erosion of mud increases steeply with ascending bed shear strength, however, but experiments show that flutes can be fashioned on strong beds at flow conditions below the plain-fluid threshold, provided sand is suspended in the flow. The grains swept along in the eddying current act as a sand blast and remove slivers from the bed as the consequence of glancing impacts. The resulting flutes have the smoothly rounded surfaces encountered on most natural marks.

Because the formation of natural flutes can be assigned from field evidence to sand-laden currents, it follows that the structures can grow only if the transported sediment is not deposited in the marks to form a protective layer. Hence the external flow must drive the separation bubble at a velocity large enough to prevent suspended grains from accumulating on the floor of the mark. We deduce from experimental data that when broad flutes are made, the flow velocity U of the current must be

$$U \geqslant U_{(crit)} = \frac{0{\cdot}91 k_2 V_0}{0{\cdot}039 k_2 - 0{\cdot}090}, \qquad (2.27)$$

in which $U_{(crit)}$ is the critical velocity of external flow for deposition of grains of free falling velocity V_0 within the mark, and k_2 is the ratio of the length to the height of the separation bubble. Fig 2.9, which shows this criterion for quartz-density grains, reveals that flutes grow only if the current is relatively fast.

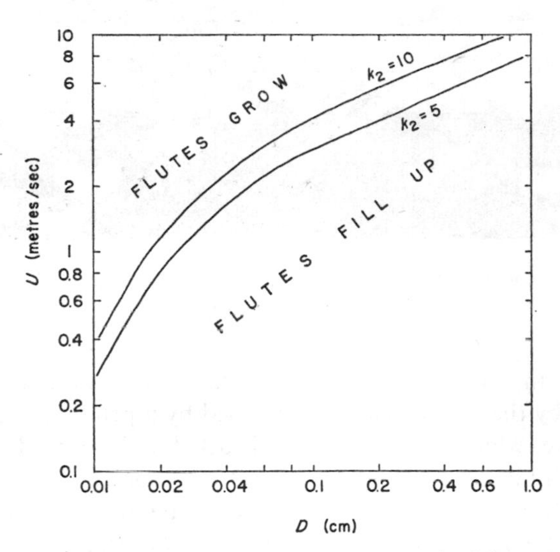

Fig 2.9 Theoretical criterion for the continuing growth of flute marks beneath a flow carrying quartz-density sand of diameter D.

It seems very probable that flutes can be initiated on a mud bed at any natural irregularity of a sufficient size and suitable shape for the occurrence of flow separation, provided that sand is not accumulating. The separation could be either fore or aft, or both together. Suitable irregularities abound on natural mud beds in the form of sedentary organisms, shells, faecal material, the entrances to living and feeding burrows, and impressed tracks and trails. The current could also create its own irregularities, in the form of chance scours, by virtue of boundary layer processes.

2.8 Load Casts and Sandstone Balls

Load casts occur typically on the bases of water-laid sandstones that overlie shales. In vertical profile they are downward swellings which vary from slight bulges to deep rounded masses (Fig 2.10). On a bedding surface, load casts are cushion-like protuberances of about the same general appearance and spacing between centres. The spacing varies from one bedding surface to another between a centimetre or two and many decimetres. Sandstone balls are ordinarily found as detached saucer-shaped to spheroidal masses

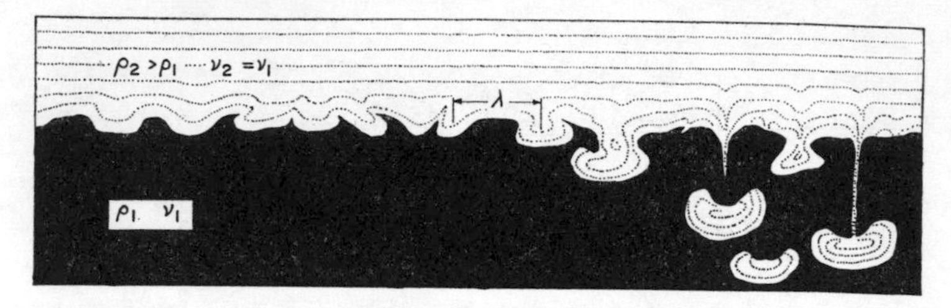

Fig 2.10 A sand bed overlying mud. At the interface are load casts passing from left to right into sand balls.

embedded in shale. Their genetic relationship to load casts is witnessed by the occasional balls attached by a pendant to the sandstone above, which may also be load-cast. Laminae inside the balls are generally bent upwards and over at the edges.

These structures represent a case of gravitational instability in which the different parts of the system in a liquified state have moved relatively prior to lithification in response to buoyancy forces. For the bulk density of fresh mud is about 1·5 g/cm^3, whereas that of sand—the upper layer—is approximately 2·0 g/cm^3. Analysis shows that the arrangement of layers is unstable to disturbances of all wave-lengths, provided the kinematic viscosities involved are neither very large nor very small. Also, for each combination of densities and viscosities, there is a disturbance wave-length λ that grows more rapidly than any other to emerge as the spacing between the centres of the cushions on a load-cast surface. With the conventions of Fig 2.10 it can be shown that

$$\frac{2\pi}{\lambda}\left(\frac{\nu^2}{g}\right)^{\frac{1}{3}} \approx 5{\cdot}2\left(\frac{\rho_2-\rho_1}{\rho_2+\rho_1}\right)^{\frac{1}{3}}, \tag{2.28}$$

for $\nu_2 = \nu_1$, and $0{\cdot}01 < (\rho_2-\rho_1)/(\rho_2+\rho_1) < 1{\cdot}0$. Thus λ decreases with ascending density difference $(\rho_2-\rho_1)$ but increases with increasing kinematic viscosity (ν).

The events following the initial condition described by Eq. (2.28) are readily demonstrated by allowing an ink drop to fall into still water from just above the free surface. The descending drop forms an inverted mushroom-shaped cloud connected to the water surface by a thin pendant, or wake, of coloured fluid. Because of drag at the surface of the descending drop, the fluid inside the drop rotates downwards and outwards about a horizontal circular axis. Similar features of shape and internal structure are to be observed from sandstone balls, which were presumably affected by drag in the same manner as the ink drop.

2.9 Sediment Sorting

Most natural sediments consist of particles ranged in size about one particular size that is represented more abundantly by grains than any other. Since the particle size distribution in most cases is also approximately log-normal, the central tendency of size is satisfactorily estimated by the logarithmic mean size, while the size spread is adequately given by the logarithmic standard deviation. The particle size spread of a sediment is its sorting, which is good if the spread of sizes is relatively narrow, but poor if the spread is comparatively broad. Experience has repeatedly shown that mean size and sorting are correlated in sands and gravels, in which the sorting worsens as mean size increases. It is also claimed that silts and clays improve in sorting as their mean size increases, but this result cannot be regarded as established, as the techniques of measurement involve severe chemical operations that would disrupt clay-mineral floccules. Very many muds are deposited in a flocculated state in fresh as well as salt water.

The mean size and sorting of the sediment at a point in an extensive deposit measure the local sorting of the deposit. Local sorting is a function of distance of transport, and this change denotes a progressive sorting. In river deposits, for example, mean size practically always decreases downstream, while sorting improves. The mean size of shallow-marine sediments decreases with increasing distance from the shore at which the sediment can be supposed to originate. However, neither local nor progressive sorting can yet be explained in unassailable terms, as we lack a sufficient knowledge of the forces that control the entrainment, transport and deposition of sediments.

Differential transport, in which particles of one size are transported faster than particles of another, probably makes a contribution to progressive sorting. It is easily seen from Ch. 1 that in aqueous transport, the mean grain transport velocity is a steeply decreasing function of grain size. Thus if a quantity of grains of mixed sizes is instantaneously released at a point into a steady flow, we shall find after the elapse of a given time that the centre of gravity of the distribution of each size on downstream distance, lies at a distance from the source proportional to the mean transport velocity. Clearly, the finest particles will on average travel the largest distance, and a progressive sorting will consequently be evident. But if the sediment supply and the transport are sustained for long enough, then the sediment flow itself becomes steady and all trace of progressive sorting due to differential transport is lost.

A more important process contributing to progressive sorting is selective entrainment. We note that natural transport systems are non-uniform, taking the large view, and that the mean local fluid force applied to the bed decreases in the direction of transport. This progressive change is observed from alluvial rivers and in moving out to sea from shore to deep water. In these situations, too, a continual local transfer of particles between bed and flow is sustained by the movement of ripples, dunes and bars and by natural fluctuations in energy level. The particles therefore move in discrete steps, between which they reside at rest in the bed. We also note from Ch. 1 that the threshold stress for entrainment of particles increases with ascending grain size. These three factors—downstream decrease of mean fluid force, continual transfer between bed and flow, and the dependence of threshold stress on grain size—act together to reduce with increasing distance of transport the probability that grains of a given size will go on being transported. Alternatively, we can say of particles of a given size that their residence time increases in the transport direction. There exists for each size of particle an 'energy fence' at which the residence time is very large, beyond which particles of this size cannot pass. This fence lies closer to the source for large particles than for some smaller size, on account of the way size determines the threshold stress, and so the deposits of the system acquire a progressive sorting.

The progressive abrasion of particles contributes negligibly to progressive sorting, except perhaps in the case of gravels of relatively soft rocks.

The local sorting of a sediment depends on the processes that control progressive sorting, inasmuch as these determine the size distribution of particles available for deposition at a site. The choice from available material is, however, made by local processes and is particularly relevant to the sorting of local samples. We can expect the sorting of sediment deposited from suspension to depend on the magnitude of the fluctuating turbulent components of fluid velocity. On the other hand, the magnitude of fluctuating intergranular forces can be expected to control the sorting of sediments deposited from a bedload layer or from a cloud of saltating grains.

2.10 Sediment Fabric

The fabric of detrital sediments is the spatial arrangement and orientation of particles. Such economically important properties of rocks as porosity and permeability depend on fabric.

The porosity of a sediment can be expressed as that fraction of the whole space which is not occupied by solid grains, while the permeability is the ease with which a fluid can pass between the assembled particles. The fractional porosity of a sediment made of regularly stacked uniform spheres varies between 0·26 and 0·48, accordingly as the spheres lie in rhombohedral or cubical packing. The number of spheres in contact with an identified sphere, called the coordination number, varies from 6 in cubical packing to 12 in rhombohedral packing. As the coordination number rises, and the passages between the spheres become narrower and more intricate, the permeability falls. Permeability is, in addition, a function of grain size, since it also depends on the actual dimensions of the passages.

The spatial arrangement of grains expressed by porosity is easily seen to depend on the magnitude of the impulse given to the bed by grains as they are deposited. The grains become more tightly packed as the impulse increases. Thus if coarse sand is slowly poured into a measuring cylinder in air, it will be observed that each grain as it strikes the bed jostles and disturbs numerous others, which settle back into relatively stable positions. The fractional porosity of the assemblage will be found to measure about 0·36. Little disturbance of grains is observed on repeating the experiment with the cylinder partly full of water, and the fractional porosity will now measure about 0·43. The impulse given to the bed by the grains as they settle in water is smaller than when they settle in air, because the falling velocity is smaller in the more viscous water. Nevertheless, large differences of fractional porosity are to be found in sediments deposited in the one medium. Sand deposited on the smooth, wave-beaten part of a beach is hard and firm beneath the feet, whereas

that laid down by ripples in a sheltered area is soft and easily disturbed, taking a deep imprint.

Sand and gravel deposits reveal a dimensional orientation of their particles which if not obvious to the unaided eye is at once visible under the microscope. The particles, in shape resembling ellipsoids, cylinders or discs rather than spheres, lie with their long axes roughly parallel to flow but inclined downwards into the current at 10–20° from the surface of accumulation. The attitude of each particle in space is the response of the particle to all the forces that acted on it during and shortly after the instant of deposition from the moving load. These forces are the body force tending to keep the particles in place, the lift force which may or may not act upwards, and the drag force tending to remove the particle from the bed. Assuming that the particle comes to rest on a bed of similar particles, the most stable position for the particle will be that when the forces acting on it are in equilibrium and when the forces of removal are at a minimum. This occurs when the particle lies with its long axis parallel to flow and tilted down at a small angle into the current. The drag force is a minimum and the lift may even be negative, while the points of contact of the particle with grains already in the bed number more than one and lie to the side and well forward of the centre of mass.

READINGS FOR CHAPTER 2

There is no real agreement as to how sedimentary structures are best classified, as some workers prefer a purely descriptive approach, whilst others favour the genetic. Somewhat contrasted views are, however, to be found in:

ALLEN, J. R. L. 1968. 'On the character and classification of bed forms.' *Geologie Mijnb.*, **47**, 173–185.

GUBLER, Y. 1966. *Essai de Nomenclature et Caractèrisation des Principales Structures Sédimentaires.* Éditions Technip, Paris, 291 pp.

POTTER, P. E. and PETTIJOHN, F. J. 1963. *Paleocurrents and Basin Analysis.* Springer-Verlag, Berlin, 296 pp.

A recent discussion of parting lineations is to be found in:

ALLEN, J. R. L. 1964. 'Primary current lineation in the Lower Old Red Sandstone (Devonian), Anglo-Welsh Basin.' *Sedimentology*, **3**, 89–108;

and the properties of the turbulent boundary layer on which the structure seems to depend are described by:

KLINE, S. J., REYNOLDS, W. C., SCHRAUB, F. H. and RUNDSTADLER, P. W. 1967. 'The structure of turbulent boundary layers.' *J. Fluid Mech.*, **30**, 741–773.

Some major features of ripples and dunes are discussed by:

ALLEN, J. R. L. 1968. *Current ripples*. North-Holland Publishing Co., Amsterdam, 433 pp;

ALLEN, J. R. L. 1969. 'On the geometry of current ripples in relation to stability of fluid flow.' *Geogr. Annlr*, **51** (Ser. A), 61–96.

HARMS, J. C. 1969. 'Hydraulic significance of some sand ripples. *Bull. geol. Soc. Am.*, **80.** 363–396.

KINDLE, E. M. 1917. 'Recent and fossil ripple-mark.' *Mus. Bull. Can. geol. Surv.*, No. 25, 121 pp.

The paper by Kindle is still a valuable source, as it has not been superseded in all respects. The relationship of current ripples and dunes to other bed forms is described by:

SIMONS, D. B., RICHARDSON, E. V. and NORDIN, C. F. 1965. 'Sedimentary structures generated by flow in alluvial channels.' *SEPM Special Publication No.* 12, 34–52;

and a host of valuable data and photographs is to be found in:

GUY, H. P., SIMONS, D. B. and RICHARDSON, E. V. 1966. 'Summary of alluvial channel data from flume experiments, 1956–61.' *Prof. Pap. U.S. geol. Surv.* 462-*I*, 96 pp.

Descriptions of sole markings, including flute marks, are profuse but amongst the best are those of:

DZULYNSKI, S. and WALTON, E. K. 1965. *Sedimentary Features of Flysch and Greywackes*. Elsevier Publishing Co., Amsterdam, 274 pp.

KUENEN, P. H. 1957. 'Sole markings of graded graywacke beds.' *J. Geol.*, **65,** 231–258.

A brief sketch of the fluid dynamics of flute marks is to be found in:

ALLEN, J. R. L. 1968. 'Flute marks and flow separation.' *Nature, Lond.*, **219,** 602–604.

ALLEN, J. R. L. 1969. 'Erosional current markings of weakly cohesive mud beds.' *J. sedim. Petrol.*, **39,** 607–623.

A review of load casts and related deformational structures is given by P. E. Potter and F. J. Pettijohn in the book already cited. The analysis of a gravitationally unstable system which led to Eq. (2.28) is afforded by:

CHANDRASEKHAR, S. 1961. *Hydrodynamic and Hydromagnetic Instability*. Clarendon Press, Oxford, 654 pp.

Three useful papers on the sorting and fabric of sediments are:

INMAN, D. L. 1949. 'Sorting of sediment in the light of fluid mechanics.' *J. sedim. Petrol.*, **19,** 51–70;

POTTER, P. E. and MAST, R. F. 1963. 'Sedimentary structures, sand shape fabrics, and permeability.' *J. Geol.*, **71,** 441–470;

REES, A. I. 1968. 'The production of preferred orientation in a concentrated dispersion of elongated and flattened grains.' *J. Geol.*, **76,** 457–465.

CHAPTER 3

Winds and their Deposits

3.1 General

Wind is the name we give to streams of atmospheric air that arise because the temperature differences between different parts of the earth produce pressure gradients. The wind experienced at a place is capable of large variations in direction, strength and gustiness. Our experience also teaches that the wind, like any other fluid in shearing motion, is able when strong enough to entrain and carry along any loose solids lying on the land surface. We see that sand is transported close to the ground, but that silt and dust are blown far up into the air and carried over large distances. Where there occurs a reduction in the power of the wind, the air-borne material is deposited to form ridge-like sand dunes or rolling dust plains. Sedimentary deposits originating in these ways are sometimes called aeolianites. They are being formed today in many parts of the globe and are known abundantly from the geological record.

3.2 Main Environments of Wind Action

The wind can actively entrain and transport loose particles only where the ground surface lacks a protective cover of vegetation and soil. These conditions are realized today in several types of region, but the three of geological interest are: (i) the deserts of the tropics, subtropics and middle latitudes, (ii) outwash plains skirting glaciers and ice-caps, chiefly in high latitudes, and (iii) sandy coasts exposed to onshore winds. The eroding and transporting activities of the wind are, of course, also favoured by climates and terrains giving rise to winds of large and sustained strength.

The occurrence of deserts in middle and low latitudes corresponds rather closely with regions of mean annual precipitation less than 30 cm (Fig 3.1). These regions occupy about 20 per cent of the total

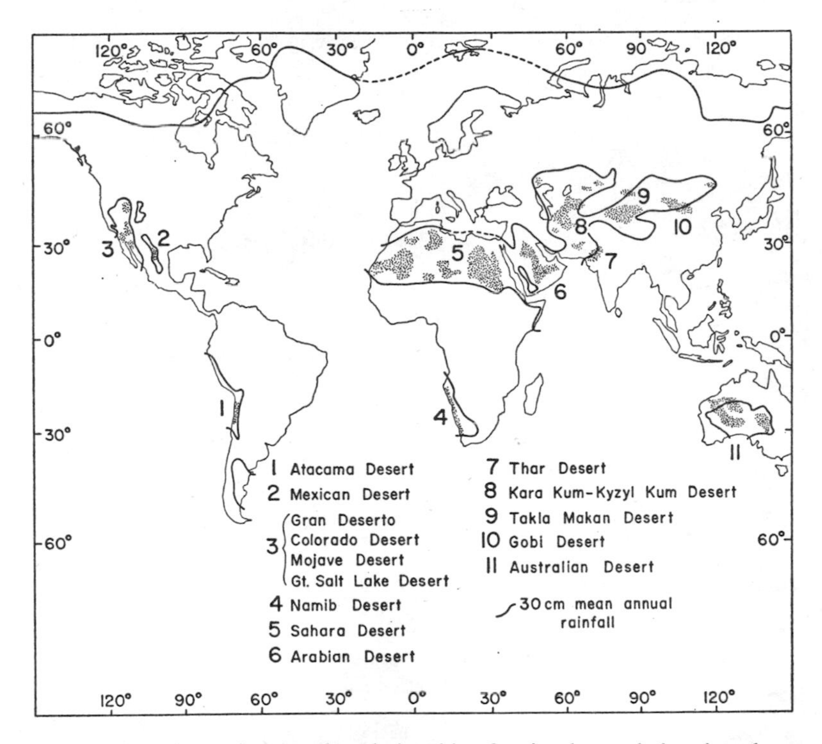

Fig 3.1 World map showing the relationship of major deserts (other than those of high latitudes) to the 30 cm mean annual isohyet. The main sand seas are stippled.

land area ($149 \times 10^6 km^2$). The largest of modern deserts are concentrated in a great belt 12000 km long and 3000 km wide rising gradually poleward between North Africa and Central Asia. The next largest desert region, measuring approximately 1500 by 3000 km, is that of central Australia. Deserts of lesser size exist in south-west Africa, Peru-Chile and Patagonia, and in south-west North America.

A distinctive association of geomorphological units is rather often

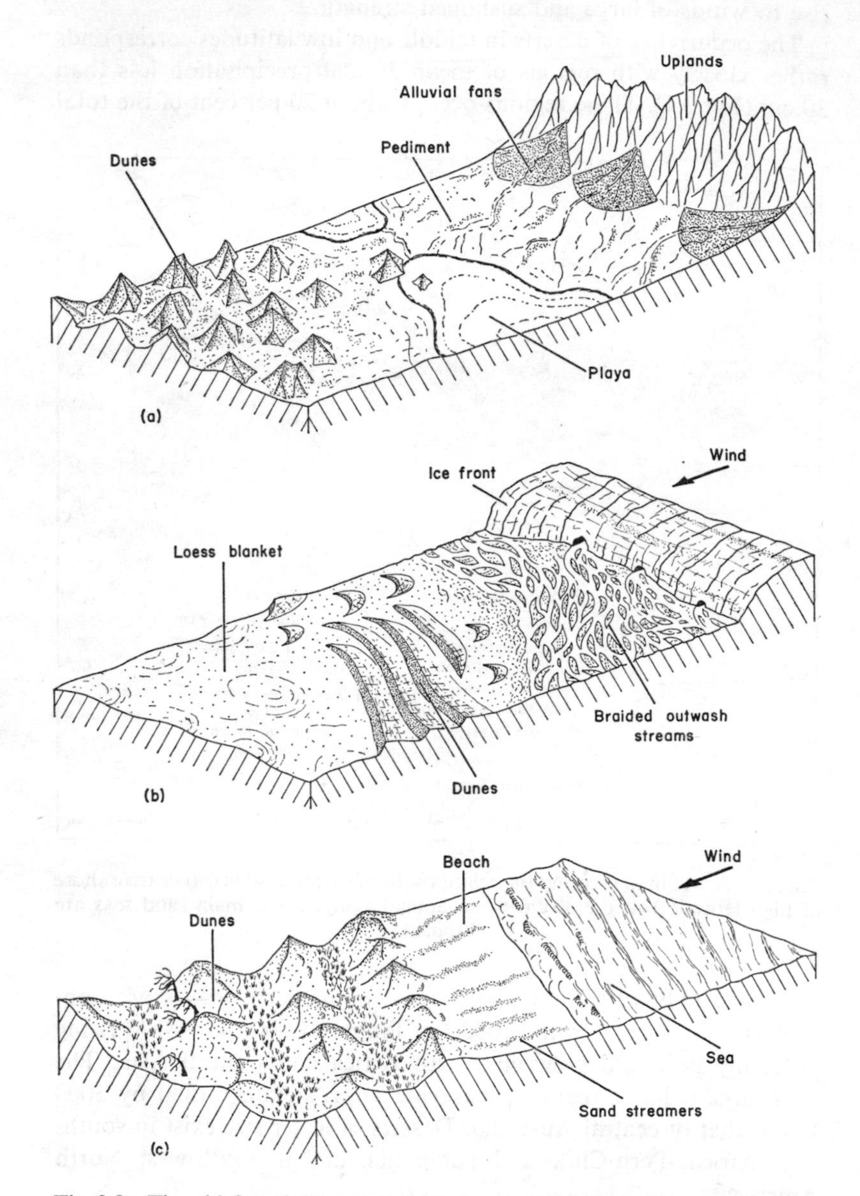

Fig 3.2 The chief environments of wind action. (a) Hot desert. (b) Glacial outwash plain. (c) Coast affected by predominantly onshore winds.

found in the deserts of low and middle latitudes (Fig 3.2a). The hills and mountains, rising steeply above the adjacent plains, are generally deeply dissected into ragged spurs, ravines and canyons. Talus-covered slopes abound. Relief forms of rounded profile are comparatively scarce and generally restricted to areas of granitic bed rock. Sloping gently away from the uplands are pediments more or less thinly and incompletely covered with alluvial material and bearing traces of the courses of ephemeral streams. The pediments in many places lead down to playas often of huge size where silt, clay and dissolved salts accumulate in temporary bodies of water. Sand dunes skirt active playas but are found as well on old playa beds and on abandoned parts of the pediment. The dunes in some areas are scattered, the old surface being visible in the gaps between them, but in other places have coalesced to form vast, deep sand seas. Locally, the dunes partly shroud rocky hills.

In many deserts, particularly the Afro-Asian ones, the units distinguished in Fig 3.2a are arranged in concentric belts, the uplands being outermost and the playas and sand seas innermost. There is in these desert basins little or no drainage from the basin to the outside; they are areas of interior drainage.

Weathering in deserts is both mechanical and chemical. The mechanical weathering depends on the generation inside exposed rocks and stones of steep, reversing temperature gradients by daytime heating and night-time cooling. Typically, the temperature at the rock surface changes by 50°C between day and night. Under such conditions the rock is violently stressed and so shatters along joint and cleavage surfaces and along the surfaces between individual mineral grains. The disruption of the rock ultimately depends, however, on a weakening of the material brought about by the chemical reactions that accompany the far from rare night-time deposition on the surface of dew or frost. Debris of a wide range of particle sizes results from combined heat-shattering and chemical weakening.

Although precipitation in deserts is low on the average, it is by overland flows of water that the debris manufactured in the uplands is flushed out on to the pediment and beyond to the playas. Sporadic but intense local rainstorms give rise in the ravines and canyons of the uplands to deep torrents that are transformed on the pediment into shallow braided streams or sheet floods. These overland flows erode the pediment, besides laying on it a deposit of loose material from which the wind can later pick at and sift the sand and dust.

The wind, now armoured with debris on account of its deflation of the waterborne sediment, contributes to the further erosion of the pediment by corrading the surface over which it travels. The importance of wind corrasion in deserts is evident from the abundance of ventifacts, polished rock surfaces, yardangs, rock mushrooms, and taffonis and arches.

Wind action is important on the outwash plains of glacial melt-water streams (Fig 3.2b), found chiefly in high latitudes, though no region of this type attains today anywhere near the size of the largest hot deserts. Again the debris ultimately deposited by the wind experiences an initial fluviatile phase of sedimentation. The winds, chiefly katabatic and cyclonic, which affect outwash plains are commonly sustained and vigorous, and blow over the plains from the ice. The chief action of the wind is to deflate the alluvial surface, which is poorly protected by plants. The eroded sand is heaped up into dunes on the plain itself or near its margins, whilst the silt is blown in great clouds over larger distances. Ventifacts are commonly abundant on outwash plains, and the rock outcrops sometimes give evidence of wind polishing.

Coasts furnished with sand by waves and tides comprise the third main situation in which the wind is important geologically (Fig 3.2c). Like a desert or an outwash plain, a sandy beach lacks a protective cover of vegetation or soil. Moreover, it consists almost wholly of material readily transported by the wind. Strong dry winds with an onshore directional component will actively deflate a moist beach which has been recently bared by the tide, and carry the dried sand inland to form dunes. Imposing dune complexes formed in this way occur along the eastern shore of the North Sea, the southern Baltic coast, the northern shores of the Gulf of Mexico, the southern shores of the Mediterranean Sea, and the west coast of Australia. Few coastal dune complexes occur in regions of low precipitation, however, and so most show a vegetation decidedly richer than encountered in deserts. As a consequence of the relatively greater protection of the surface, the dunes of coastal complexes are usually loosely stabilized and of irregular form.

3.3 Entrainment and Transport of Sediment by Wind

A stationary particle resting on a flat surface of similar grains can be entrained by the wind in one of two ways: either the particle is dislodged by the application of a sufficiently large fluid force, or it is set in motion by being struck by a grain already in flight. Entrainment by the first method is amenable to an analysis on lines already indicated in Ch. 1. In the case of wind, it is convenient to write

$$U_{*(\text{crit})} = k_1\sqrt{\frac{\tau_{0(\text{crit})}}{\rho}} = k_1\sqrt{\left\{\frac{(\sigma-\rho)}{\rho}gD\right\}}, \tag{3.1}$$

in which $U_{*(\text{crit})}$ and $\tau_{0(\text{crit})}$ are the shear velocity and shear stress, respectively, at the threshold of movement, σ is the density of the grains, D is the diameter of the grains, ρ is the fluid density, and k_1 is a constant approximately equal to 0·1 provided the grain Reynolds number $Re = U_*D/\nu > 3{\cdot}5$. But if the Reynolds number falls below 3·5, the boundary becomes aerodynamically smooth and k_1 increases as a function of the declining Reynolds number. In this range of conditions the square root law of Eq. (3.1) no longer holds. As plotted in Fig 3.3 the curve of $U_{*(\text{crit})}$ against $\sqrt{(\sigma D)}$ shows a minimum at a diameter for quartz-density solids of about 0·10 mm. Material of this size is therefore the easiest to set in motion. The erosion of fine silt by the wind requires the application of about the same force as the entrainment of coarse sand, a fact which is readily demonstrated by attempting to blow particles off a smoothed surface

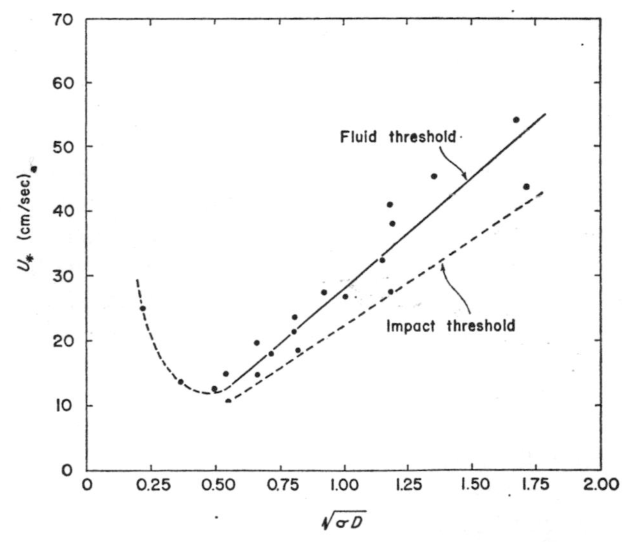

Fig 3.3 Experimental thresholds of movement for quartz-density sand under wind action (data of K. Horikowa and H. W. Shen). For convenience D is given in millimetres.

G

of baking flour or portland cement. Experiments have confirmed the validity of Eq. (3.1) in the rough boundary range, and have given some support to the existence of the inflection point.

The solid curve of Fig 3.3 defines what is called the fluid threshold for particle entrainment by wind. A lower threshold value of U_*, the impact threshold, is found if the wind bears with it particles already eroded from the bed. For $U_*D/\nu > 3{\cdot}5$, the impact threshold value of U_* appears to be about 80 per cent of the fluid threshold. The impact threshold is very much smaller than the fluid threshold for a bed of silt beneath a flow bearing sand, for the sand grains on landing throw up the silt from small impact craters. Similar factors favour the erosion of dust where animal or vehicle traffic crosses a dried surface.

The wind transports sediment in suspension, saltation or surface creep. Silt is ordinarily transported in suspension, whereas sand moves by a combination of creep and saltation.

Saltating sand grains proceed by a bouncing motion initiated by granular impact (Fig 3.4). Wind tunnel observations show that the

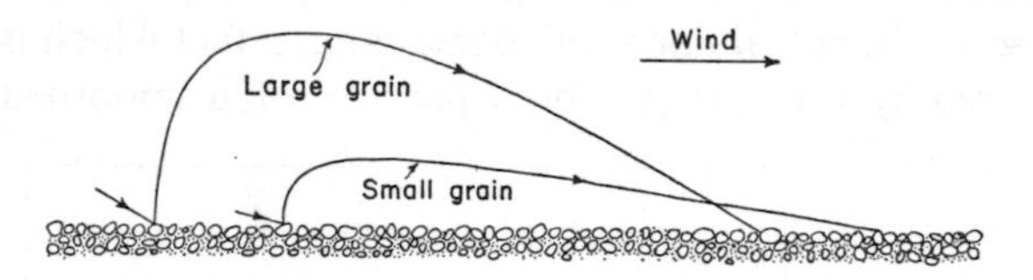

Fig 3.4 Trajectories of saltating sand grains.

grains travel over low trajectories at velocities parallel to the ground of the order of the wind speed. The initial rise of a saltating grain is directed almost vertically up from the bed, but the later part of the trajectory is long and flat, and the grain returns to the bed at an angle seldom greater than 10° and rarely smaller than 3°. On striking the bed, the grain may either bounce up again off a stationary particle, thus regaining the higher parts of the wind flow where its previous momentum can be restored, or, sinking into the bed, may splash up grains which in their turn ascend vertically into the wind. Theoretically, the mean length and height of the grain trajectory during movement over a continuous sand bed under a constant wind of shear velocity U_* become

$$L = k_2 \frac{(U_* + U_{*(\text{crit})})^2}{g}, \tag{3.2}$$

and

$$H = k_3 \frac{(U_* + U_{*(crit)})^2}{g}, \tag{3.3}$$

in which L and H are the mean length and height, respectively, of the trajectory, k_2 and k_3 are dimensionless empirical constants depending on grain size, and g is the acceleration due to gravity. The constant k_2 appears insensitive to grain size and is 0·65 for a fine grained sand. It follows that L increases steeply with ascending U_* but is little influenced by grain size. It appears from theoretical considerations and field observation that k_3 increases rapidly with ascending grain size. It is, however, of the order 0·05 for a fine sand.

The surface creep comprises those grains which, instead of being ejected upward at a steep angle into the wind, travel slowly and unsteadily along the surface under the bombardment of the saltating particles. It is to be expected that a high proportion of the coarser particles in a load of mixed sizes will travel in the surface creep, though individual grains of any size will alternate between creep and saltation as the conditions change from one impact to the next.

The relationship between the sand transport rate and the force exerted by the wind is a matter of great importance. Two theoretical solutions, one by R. A. Bagnold and the other by R. Kawamura, are available, together with several empirical transport functions of less obvious value.

Bagnold argues that the transport of sand denotes a continual loss of momentum by the wind. If a quantity i_s of sand in saltation moves along a lane of unit width and passes a fixed point in one second, the rate of loss of momentum by the air is

$$i_s \frac{(u_2 - u_1)}{L}, \tag{3.4}$$

per second per unit area of surface, where u_1 is the average horizontal velocity with which the grains rise from the ground and u_2 is the average horizontal velocity at which they travel over a trajectory of average length L. Neglecting u_1 as small compared with u_2, we equate the rate of loss of momentum to the drag force exerted on the wind, and obtain

$$i_s \frac{u_2}{L} = \tau_0 = \rho U_*^2. \tag{3.5}$$

But u_2/L is found to approximate very closely to g/v over a wide range of conditions, where v is the initial vertical velocity of the grains at the start of their trajectories. Arbitrarily assuming that $v = k_4 U_*$, where k_4 is an impact coefficient, we deduce that

$$i_s = k_4 \frac{\rho}{g} U_*^3. \tag{3.6}$$

In order to include the rate of sand transport by surface creep, estimated by Bagnold to be about 25 per cent of i_s, together with the effects of particle size, Eq. (3.6) may be rewritten as

$$i = k_5 \left(\frac{D}{D_{st}}\right)^{\frac{1}{2}} \frac{\rho}{g} U_*^3, \tag{3.7}$$

in which i is the total sand transport rate, D_{st} is the grain diameter of a standard 0·25 mm sand, and k_5 is an empirical coefficient depending only on the sorting of the load ($k_5 = 1{\cdot}8$ for natural sands).

Kawamura postulates that the shear stress τ_0 exerted by the wind consists of τ_s due to the impact of sand grains and τ_w caused by the wind directly. Then

$$\tau_0 = \tau_s + \tau_w. \tag{3.8}$$

But since under conditions of steady sand flow, τ_W is found to equal $\tau_{0(crit)}$, we can write

$$\tau_0 - \tau_{0(crit)} = \tau_s = W\overline{(u_2 - u_1)}, \tag{3.9}$$

where W is the mass of sand falling in unit time over unit area and the bar above the bracket denotes the mean value. Substituting vertical for horizontal velocities in a form defined by the shear velocity of the wind, and introducing L from Eq. (3.2), the final expression for the transport rate can be shown to become

$$i = WL, \tag{3.10}$$

which can be expanded to

$$i = k_6 \frac{\rho}{g} (U_* - U_{*(crit)}) (U_* + U_{*(crit)})^2, \tag{3.11}$$

where k_6 is a constant to be obtained experimentally.

It will be seen that Eqs. (3.7) and (3.11) are virtually identical at large transport rates but that they deviate increasingly as the shear velocity decreases. Bagnold's function states that $i \neq 0$ when $U_* = U_{*(crit)}$, whereas Kawamura's yields the more realistic result that $i \to 0$ as $U_* \to U_{*(crit)}$. Both equations are fairly well substantiated by field and wind tunnel measurements, though there remain uncertainties regarding the coefficients. There is evidence that particle shape has a strong influence on the bouncing properties of grains and hence the flatness of their trajectories.

The above transport functions depend on momentum considerations and not on energetics as favoured in Ch. 1. This is only because it is difficult to specify the power of a wind, which has no easily definable thickness and flows under the influence of a pressure gradient. Both transport functions state that the transport rate is proportional to the 3/2 power of the boundary shear stress, and the same general result will be deduced from considerations of work in the case of transport by rivers (see Ch. 4).

Experience teaches that the density of the sand cloud driven along by the wind rapidly decreases upwards from the surface. If W is the mass of sand passing in one second through unit area measured in a vertical plane normal to the wind at a distance y from the bed, then the exponential equation

$$W = k_7 \rho U_* \exp\left(-\frac{4{\cdot}3\sqrt{(gy)}}{U_*}\right) \tag{3.12}$$

(using the notation $\exp(x)$ to mean e^x, e being the base of natural logarithms) approximates very closely to the distribution of W with y. The dimensionless constant k_7 depends on grain size.

Grain size at first decreases and then increases within the lower part of the driven sand cloud, as can be seen from the experimental results given in Fig 3.5. The point of reversal of the trend lies within a few centimetres of the bed and increases in height above the bed with ascending U_*. Field measurements show that above this point the sand in saltation goes on increasing in size over a normal distance of several decimetres or metres above the bed. Over larger distances from the bed the grain size of sand in the cloud decreases. Hence the coarsest transported sand is concentrated at two levels: at the bed itself, and at a considerable distance upwards from the bed. This complex vertical distribution of sizes is explained by the fact that although transport in the surface creep is typical of large grains, when such grains travel by saltation they take trajectories that are higher on average than the finer particles (see Eq. 3.3). However,

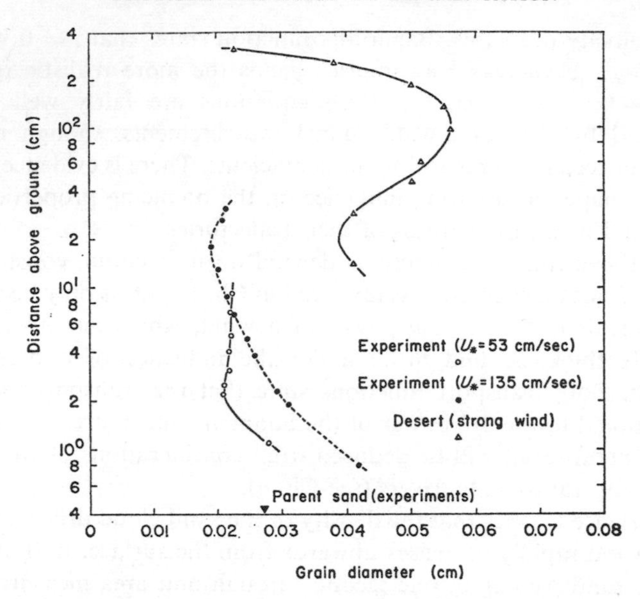

Fig 3.5 Experimental and field data on average diameter of wind-driven sand as a function of distance above undisturbed sand surface (data of T. Y. Chiu and R. P. Sharp).

the transport rate of coarse particles at the upper level is insubstantial.

The transport of silt in turbulent suspension by the wind is difficult to treat theoretically because knowledge is required of the gross structure and thermal stability of the winds concerned. Silt and dust incorporated into an advancing front, for example, will attain great heights rather quickly on account of the upward convective movement of air at the nose of the front. However, assuming air that is neutrally stable thermally, and no special circulation pattern, the diffusion equation discussed in Ch. 1 permits us to write for the vertical concentration of silt particles of free falling velocity V_0

$$\frac{C_y}{C_a} = \left(\frac{y}{a}\right)^{-V_0/\kappa U_*}, \tag{3.13}$$

where C_y is the concentration at a height y, C_a is the concentration at a reference height a, and κ is the von Kármán constant assumed equal to 0·4. If the silt is eroded from a narrow strip of ground extending for a large distance across the wind, it can be shown that the height of the dust cloud increases with distance downwind as

$x^{7/9}$, where x is the downwind distance, while the silt concentration close to the ground varies as $x^{-8/9}$. But these simple relationships are progressively altered as the air mass departs from neutral stability. If the stability increases, the vertical concentration gradient is progressively steepened relative to the neutral condition. The dust cloud is not uncommonly restricted to the first few metres above the ground. In unstable air, in which buoyant masses are rising locally, the vertical concentration gradient is lower relative to the neutral condition because of the additive effects of convection. As in the case of diffusion with convection at an advancing front, the dust cloud may rise rapidly to a great height, of hundreds or even thousands of metres, and subsequently spread its load over a wide area.

3.4 Textural Features of Wind Deposits

The processes of transport and deposition in air sufficiently resemble those in water as to lead to a general textural similarity between wind-laid and water-laid sediments. Besides, dune sands and loess are chiefly derived from fluviatile or littoral parents, and it would be surprising if all traces of their origin were obliterated by wind action. Such differences as can be observed between sediments laid down in the two media are all subtle, depending for recognition on a statistical evaluation of large numbers of carefully collected and analysed samples.

Wind-blown dune sands are chiefly unimodal in size distribution, with a mean size seldom less than 0·20 mm and rarely greater than 0·45 mm. On the average, dune sands are significantly better sorted than sands collected from river beds, though there is a substantial overlap of the sorting-mean size fields of the two groups of sands. However, the sorting of river sands worsens with increasing mean size, while no such trend has yet been detected in the case of wind-blown dune sands. The latter are further typified by a small to moderate positive skewness, i.e. their size frequency curves slope most gently on the fine grained side of the plot. River sands, on the other hand, range very widely in skewness, from strongly positive (much more so than dune sands) to strongly negative. Size for size, beach and dune sands are about equally well sorted. Beach sands are, however, negatively skewed and often strongly so.

The deposits underlying wind-deflated areas, between dunes in sand seas and on the borders of sand seas, are generally poorly sorted and commonly bimodal. Usually the coarser mode occurs at a grain diameter of 0·5–1·0 mm, while the finer mode, which frequently is the dominant one, lies at a diameter of about 0·10 mm. The sizes represented above the coarser mode generally include small pebbles. However, the range of sizes bounded by the two modes is broadly that representative of the dune sands, indicating that the inter-dune areas experience a preferential deflation; the material of the coarser mode is ordinarily too big to be moved by the wind, whereas that of the finer mode is readily trapped in the hoppers formed by the larger grains.

The ease with which silt is maintained in turbulent suspension in the air, combined with the readiness with which such materials can be eroded by a sand-bearing wind, means that the wind is highly effective in separating silt particles from sand which can travel only close to the ground and on average at a speed well below that of the wind. Wind-blown silt, or loess, forms a blanketing deposit to a thickness commonly of several metres over huge tracts of country surrounding hot deserts and bordering river valleys and outwash plains in periglacial regions. Some wind-blown silt even contributes to ocean-bed sediments, having been blown thousands of kilometres from the desert by such strong seasonal winds as the northerly *harmattan* of West Africa. On land the thickness and mean grain size of loess deposits ordinarily decreases exponentially with distance from the source of the silt. Loess deposits have a mean grain size seldom less than 0·015 mm and rarely greater than 0·045 mm. They are well to very well sorted and on average significantly better sorted than water-laid silts whose clay content is generally substantial. Amongst the other distinguishing features of loess deposits are lack of laminations, the occurrence of fossil soil horizons, and an often rich fauna of land snails.

3.5 Desert Dunes and Sand Seas

The wind in its course heaps up sand into a great variety of accumulations classifiable as either wind ripples or ridges, drifts, or dunes. Wind ripples are the smallest transverse mounds into which sand grains can be heaped; ridges are somewhat larger than ripples, but are not as large as dunes, and commonly smaller than drifts. Accumulations caused by an obstruction—bushes, rocks or cliffs—in the path of a sand-laden wind are called drifts. On the other hand, dunes are sand accumulations able to move freely under the shifting wind, and are ordinarily found in dense association with other accumulations of the same kind. These associations contribute to sand seas. An individual dune may be completely isolated from its neighbours upon a pavement of rock or stony desert, or it may join up with them along its lower slopes, the surface consisting everywhere of blown sand. We know little of the factors that determine the characters of dunes and drifts, chiefly because of the lack of adequate long-period observations of wind strength and direction in deserts, and partly because of difficulties of close examination under sand-driving conditions. The controlling factors of greatest importance appear, however, to be the availability of transportable material and the strength and directional consistency of the wind. Only effective, i.e. sand-driving, winds are of interest here. Reference to Eqs. (3.7) and (3.11) will show that the transporting ability of the wind increases very rapidly with increasing wind gradient.

Wind ripples abound on surfaces of blown sand. They are laterally extensive, straight to slightly sinuous crested transverse grain ridges of wavelength seldom less than 2 cm and rarely greater than 15 cm. The ripples are slightly asymmetrical in profile, the gentler slope lying to windward, and of height between 0·4 and 1·2 cm. It appears that wind ripples arise when a flat surface across which sand is blown is unstable. The local sand-removing action of the cloud of sand in saltation is accentuated by any chance deformation of the surface, being greater on slopes facing into the wind than on slopes sheltered from the wind. This being the case, the wavelength of wind ripples should be equal to the mean length of the sand grain trajectory, as given by Eq. (3.2), and should increase with ascending wind strength as measured by U_*. It is found, in the open and in wind tunnels, that there is a fair agreement between the measured and calculated wavelengths of wind ripples.

Wind ridges, sometimes called granule ripples, are transverse structures larger than wind ripples having wavelengths generally between 0·25 and 3·0 metres and heights of 2·5 to 15 cm. The ridges consist of coarse sand, granules and small pebbles, all of which are too massive to move by saltation in even the strongest natural winds, but are not so big that they will not move by creep under the continual bombardment of finer particles. The ridges are therefore chiefly found in deflated areas.

Sand accumulates in smooth, laterally extensive rising drifts against moderately inclined cliffs that face into the wind (Fig 3.6a). The drifts occur in response to a reduction of the sand-transporting capacity of the wind consequent on an incipient or weak separation of the wind forward of the barrier. Commonly such drifts rise to a height of several hundred metres above the adjacent plains, and those of the Peruvian desert are particularly large. Drifts to leeward of cliffs and rock ridges, however, often include a series of sharp crested sand ridges that taper down-wind from the obstacle (Fig 3.6b). The ridges, with analogues amongst water-shaped bed forms, are a result of the action in the lee of the cliff or ridge of large vortices with axes parallel to the general wind flow. The drifts that form in the lee of permeable bushes or grass tufts chiefly are tapering wedges of sand marked by a sharp crest and a pattern of wind ripples

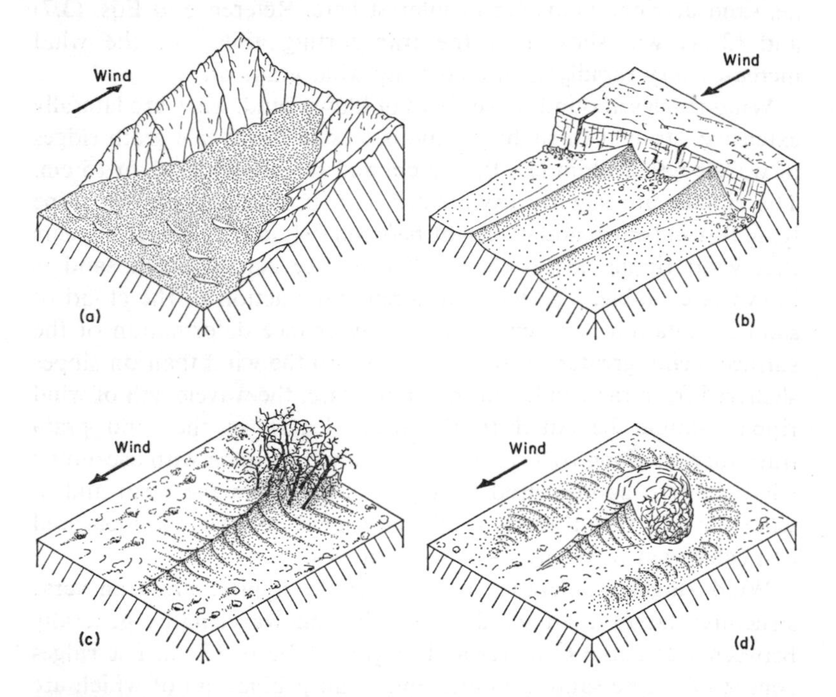

Fig 3.6 Idealized sand drifts formed by wind. (a) Rising drift upwind of tall cliffs. (b) Ridges to lee of low cliff. (c) Drift in lee of permeable bush. (d) Drifts around and in lee of isolated boulder.

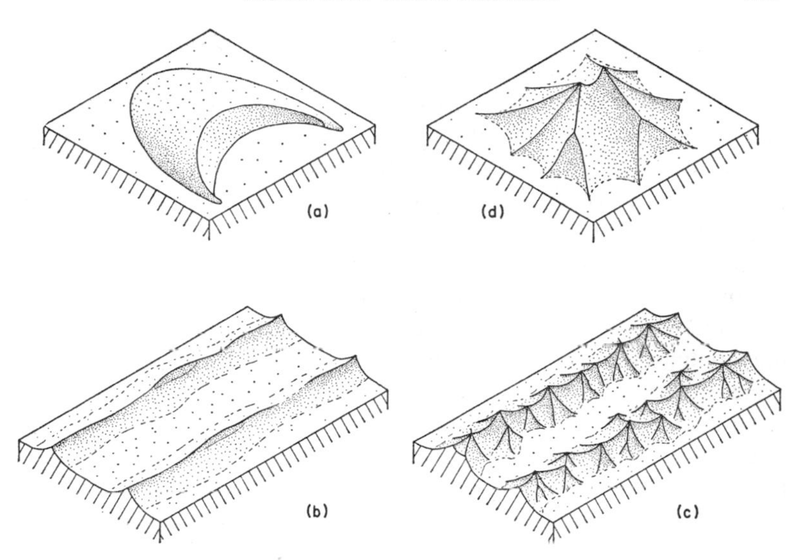

Fig 3.7 Idealized sand dune types. (a) Barkhan. (b) Simple longitudinal. (c) Complex longitudinal. (d) Pyramidal.

denoting movement of grains upwards towards the crest (Fig 3.6c). A boulder or comparably shaped bluff body gives rise on the open desert surface to a more complicated drift, marked by a crescentic ridge of sand to the front and sides of the body in addition to a trailing ridge (Fig 3.6d). In such cases the sand accumulates in response to forward as well as rearward separation of the wind flow.

Sand dunes in deserts assume many different, probably intergrading, forms, but can be divided between three principal types according to whether the longest dimension lies parallel to or across the wind, or whether the structure has no preferred dimensional orientation.

Amongst transverse dunes the best known is the crescent-shaped barkhan (Fig 3.7a) which ranges in height between 0·5 and 40 metres, in length parallel to the wind between 2·5 and 250 metres, and in width across the horns between 5·0 and 400 metres. Sand leaves a barkhan dune chiefly at the tips of the horns, and it appears that the constant shape of the dune represents a state of equilibrium between one pattern of air and sand flow tending to elongate the horns and another, different pattern striving to shorten them and straighten the dune crest. The most perfectly shaped barkhans occur as completely isolated structures on rock or stony surfaces where the amount of transportable sediment is small. As the availability of sand increases, the dunes join up at the horns with a consequent lengthening and decrease of curvature of the crest. When the stage of a continuous sand cover is reached, the dunes are found to have slightly sinuous crests several tens or hundreds of times longer than the dune wavelength, commonly between 50 and 200 metres. The troughs between the dunes are divided up across the wind into deep hollows by sharp crested sand ridges that project down-wind from the slip faces. These spurs and hollows are shaped by a large-scale and complex motion of the air in the lee of each slip face. Transverse dunes appear to depend on a regime in which effective winds blow from a narrow fan of directions.

Longitudinal dunes lie parallel to the wind at a rather constant transverse spacing. The smallest and simplest (Fig 3.7b), as encountered for example in the Australian desert, are between 5 and 15 metres high and between 200 and 2000 metres transversely apart. They commonly run for 50 kilometres without a break and join up or part at Y-shaped junctures. The dunes stand on a deflated surface varying from stony to clayey and have plinths lightly dotted with scrub. The beaded appearance of the dunes in plan, and their saw-tooth elevation, reflect the alternation of summits and saddles along the dune length. At the summits are low slip faces that change direction as the wind blows now from one side then the other of the dune. The largest and most complex of longitudinal dunes, as found in parts of the Sahara and Saudi Arabian deserts, range in height to more than 200 metres and in transverse spacing to upwards of 10 kilometres. They commonly extend for several hundred kilometres without a break in the crest and consist of a linear complex of smaller, somewhat irregular dunes and sand waves amongst which transverse and equidimensional types are recognizable (Fig 3.7c). The broad lanes between the dunes are deflated areas underlain by rock, stony material, poorly sorted sand and silt, or playa clay. We are still unable to account satisfactorily for the regular spacing and persistence parallel to the wind of longitudinal dunes. The evidence in some cases suggests that the dunes denote the resultant between two equally effective winds lying about 90° apart in direction but blowing at different times of the day or year. Elsewhere winds of small to moderate strength blowing from a narrow range of directions are in evidence, and the dunes appear to form in response to secondary currents set up in these winds. Of possible importance in this connection may be the thermal instability of the air heated by contact with the desert surface.

Two other varieties of longitudinal dune are locally important. That called sinuous is represented by relatively short, simple, isolated ridges in plan wavering slightly from side to side. The second, hook-shaped variety is represented by lines of ridges each shaped like small *mu* in Greek, the tail of one ridge fitting into the cup at the head of its predecessor.

Where effective desert winds box the compass, we find equidimensional dunes of which the commonest are those called pyramidal or star-shaped. Dunes of this type are characterized by steep, radial, sometimes branching *arêtes* of sand that culminate upwards in one or more pointed summits (Fig 3.7d). The smaller dunes have but a single peak and are generally a few hundred metres in diameter and between 10 and 30 metres in height. The larger dunes, with several summits, are 1000 to 2000 metres across the base and up to 250 metres in height. Very commonly, pyramidal dunes form large fields, the members of which are arranged at a remarkably regular spacing apart. The ground between these dunes is deflated and often reveals the basement material. Dome-shaped dunes, another variety of equidimensional dune, are apparently restricted to the Saudi Arabian deserts where they are widely distributed. The dunes are circular to oval in plan and between 800 and 2000 metres across the base. Their flat tops bear small dunes of transverse types, though on the tall sides are sand *arêtes* comparing with those of pyramidal dunes.

We may finally mention what are best called complex dunes. These are enormous isolated sand hills often more than 150 metres in height and 5 kilometres across the base. In plan they vary from oval to triangular, and some resemble barkhans but on a huge scale. Complex dunes have enormously tall slip faces on the leeward side, while the windward slopes bear transverse dunes of conventional size and shape.

Although sand seas occur in nearly every major desert region, few are at all well known as regards the type, distribution and orientation of the dunes found in them. A notable exception is the Australian desert (Fig 3.8), where the sand seas consist of longitudinal dunes of simple form whose orientation in relation to wind regime is known in detail. A more complex sand sea is the Erg Oriental in the

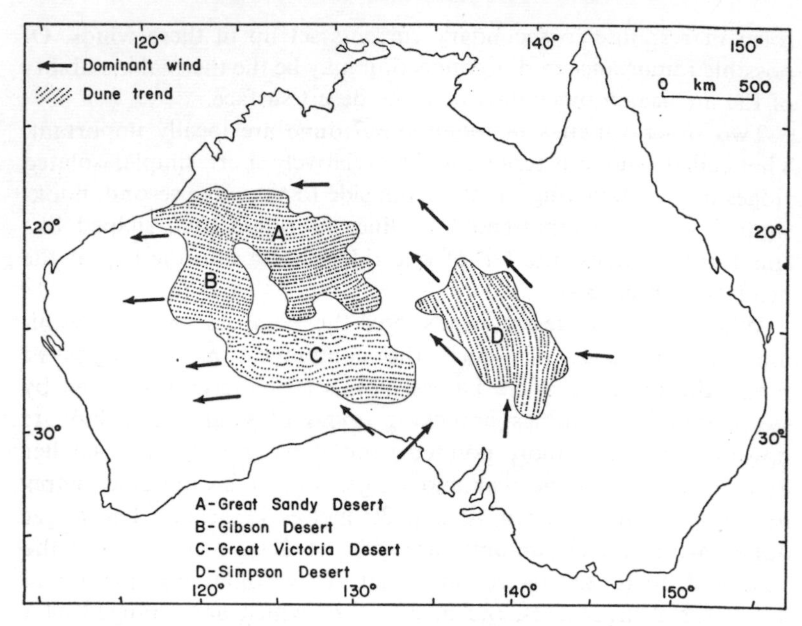

Fig 3.8 The Australian sand seas (data of C. T. Madigan).

Algerian Sahara (Fig 3.9). The peripheral areas of this sea comprise longitudinal dunes, whereas in the central zone are chiefly equi-dimensional dunes. The wind regime of the Erg Oriental is, however, a good deal more complex than that of the Australian desert.

3.6 Coastal Dunes

Wind-blown dunes are commonly found to surmount beach-ridge plains on coasts subject to strong onshore winds. However, it is only in dry regions, where vegetation is negligible, that such dunes attain regular form. Long-crested transverse dunes are the commonest coastal dunes of arid and semi-arid regions, though barkhans and barkhan-like forms occur where deflation exposes the water-table. The dunes increase progressively in height inland from the beach providing the sand, but the crests seldom exceed a height of 5 metres above the troughs. The shape of coastal dunes in wetter regions is largely controlled by plants whose dual role is to promote the deposition of sand by the wind and to anchor the sand once deposited.

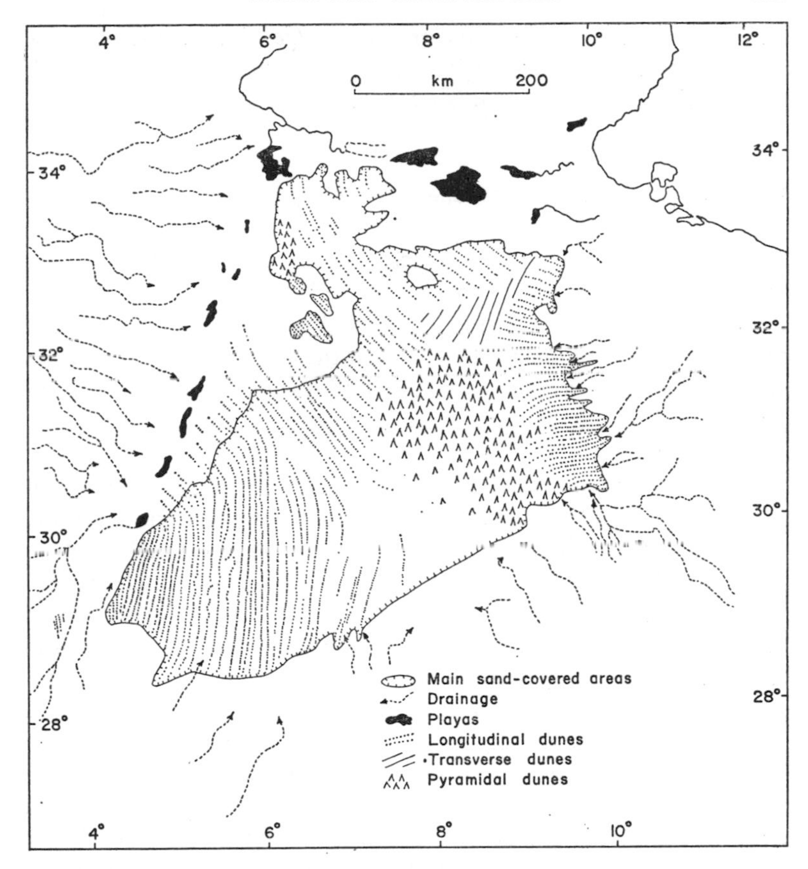

Fig 3.9 The Erg Oriental in the Algerian Sahara (data of L. Aufrère).

The resulting dunes are commonly pyramid-like but quite irregular in size and shape, having none of the *arêtes* so characteristic of equidimensional dunes in the desert. Moreover, they are small and practically never taller than 20 metres. Parabolic or blowout dunes are frequently found amongst the coastal dunes of temperate and humid regions, and in some complexes are the predominant form. A parabolic dune consists of a deep, central hollow elongated parallel to the wind and closed off to the sides and at the down-wind end by a sharp sand ridge whose outer surface, though partly vegetated, is a simple or composite slip face. Thus, in orientation relative to the wind the parabolic dune is the opposite of the barkhan. Parabolic dunes vary greatly in size, but lengths of a few hundreds of metres and heights of a few metres are fairly typical. The dunes are two to four times longer than wide.

3.7 Internal Structure of Dunes

We know little of the internal structures of dunes, though much may be guessed at, as natural exposures of the interiors of modern dunes are rare and practical considerations seldom allow an investigation deeper than the first metre or two below the surface. Large quantities of water to stabilize the sand, and bull-dozers to excavate it, are called for, and neither are readily made available in the desert regions of interest.

The stratification of wind-blown sand involves two types of laminae and bounding surface between sets of laminae. The bounding surfaces are easily dealt with; they are either erosional or non-depositional. Avalanche or foreset laminae consist of loosely packed sand and are formed when sand slides under gravity down the slip face in the lee of dunes. The laminae are inclined to the horizontal between 28° and 42°, depending on the material, and are straight except near their bases where they approach the underlying surface tangentially. A lateral sequence of avalanche laminae in a dune makes up a cross-bedded unit whose constituent foresets denote the successive earlier positions of the dune slip face, as we have seen in Ch. 2. Accretion laminae, on the other hand, consist of tightly packed grains deposited directly from the creep and saltation loads of the wind. These laminae are convex-upward, straight, or concave-upward, and dip at angles chiefly smaller than 10°, although dips as high as 25° are not uncommon. Accretion laminae are formed chiefly on the windward slopes of dunes, though some arise laterally or to lee.

The internal cross-bedding of a large transverse dune excavated to its base is shown in Fig 3.10a. This dune built upward as well as forward, and there was from time to time erosion on the leeward side, which led to rounding of the dune crest. The trench cut parallel to the dune crest reveals shallow, trough-like structures indicating the sinuous plan of the crest and slip face. The avalanche laminae evidently dip within a single narrow fan of directions. Other transverse dunes have been excavated which show cross-bedding units of opposing dip, indicating reversal of the slip face under a changeable wind.

As can be deduced from geometrical considerations, the internal structure of a simple longitudinal dune is characterized by cross-bedded units whose dip azimuths lie nearly 180° apart symmetrically about the elongation direction of the dune crest. This inference is confirmed by the structures observed in shallow pits cut in a small

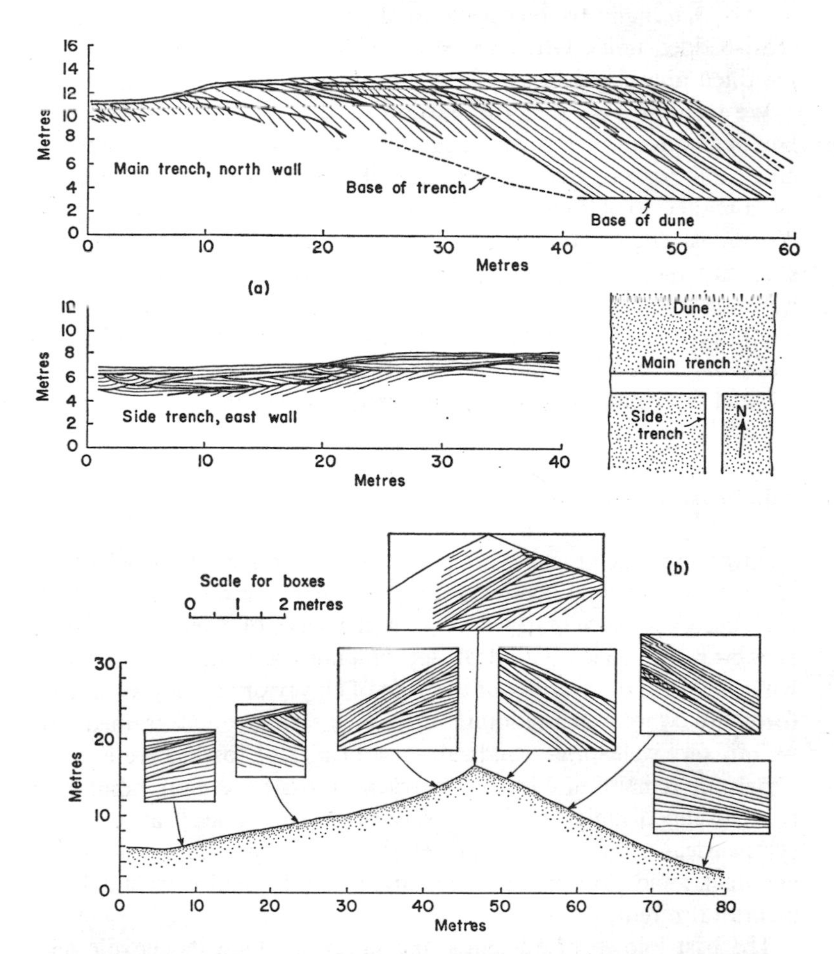

Fig 3.10 Internal bedding structure of wind-blown sand dunes. (a) Transverse dune, White Sands National Monument, New Mexico (after E. D. McKee). (b) Transverse section of simple longitudinal dune, Libyan Sahara (after E. D. McKee and G. C. Tibbitts).

longitudinal dune about 80 metres wide across the base (Fig 3.10b). It will be noticed that laminae with the steepness of avalanche laminae are confined to the upper slopes of the dune, whereas laminae with the low dips of accretion laminae are dominant in the plinth. Longitudinal dunes of the kind sketched in Fig 3.7c would undoubtedly have a much more complicated internal structure.

The irregular and partly stabilized dunes of coastal complexes are marked internally by relatively small, cross-cutting, trough-shaped cross-bedded units. Often the surfaces bounding these units below are tilted downward in the direction of the prevailing wind.

We would expect that the pattern of dip azimuths of avalanche laminae mapped over a whole sand sea was controlled by dune type and wind regime. If the dunes are of the transverse type and the wind regular, the dip azimuths would map within a narrow range of directions. The variance of the azimuthal directions would reflect the sinuosity of the dune crest and the variability of the wind. If, however, the dunes are of the longitudinal type, the dip azimuths would plot in a bimodal distribution bisected by the mean effective wind direction. A sea of equidimensional dunes would be expected to yield dip azimuths without preferred orientation.

3.8 Ancient Wind Deposits

Contrary to legend, no wholly reliable criteria exist by which to identify ancient wind deposits, and the problem of reliable identification becomes particularly severe in the case of wind-blown silts. However, the following ideal set of characters strongly suggests wind-blown sand: scarcity or absence of clayey or gravelly sediment; formation wholly or predominantly of fine to medium grained and well to very well sorted sand; absence of marine fossils; presence of vertebrate remains and especially tracks, cross-bedding in tabular or trough-shaped units typically several metres thick; absence of symmetrical ripple mark; presence of rain prints; presence of ventifacts; and close spatial association with formations of fluviatile or littoral origin.

The best known of ancient dune sands are Late Palaeozoic and Mesozoic in age. In Britain, dune sandstones of Permian age crop out along the Moray Firth, in the Southern Uplands of Scotland and the Lake District, in Yorkshire, Nottinghamshire and the West

Midlands, and in Devon. The formations concerned reach a maximum thickness of about 260 metres and rest on an irregular surface of older rocks. They consist of well sorted fine to medium sand and are cross-bedded throughout in wedge or trough-shaped units mostly between 2 and 8 metres thick. That the dunes of these ancient sand seas had curved crests is shown by the winding of avalanche laminae exposed on surfaces parallel to the general surface of accumulation of the time. It has been claimed that the Permian dunes were barkhans, though it is more likely that they were long-crested transverse types. The azimuth directions of avalanche laminae have been mapped in the outcrops mentioned above, and the results indicate that the Permian wind blew consistently from what is now east. The Permian aeolian deposits of Great Britain pass laterally into marine strata and are succeeded by fluviatile sediments. In a few localities the dune sandstones interdigitate with what appear to be fluviatile deposits.

In the Colorado Plateau region of the USA hot desert conditions prevailed throughout Late Palaeozoic and Mesozoic times, so that there now exists in this area a whole series of aeolian sandstone formations closely associated vertically and horizontally with fluviatile and shallow-marine strata. The Colorado Plateau must during these times have closely resembled a modern desert basin of interior drainage, for it was bounded to west, south and east by hills and mountains, and was open to riverine lowlands and the sea only to the north. The most closely studied of the dune sandstones are the Coconino and the Navajo, neither of which is thicker than about 370 metres. Their cross-bedded units are seldom thinner than 1·5 metres and practically never thicker than 12 metres, though exceptional units 30 metres thick are recorded. The sands are well sorted and here and there yield ventifacts. Vertebrate tracks, rain prints and undoubted wind ripples are recorded from the Coconino. The winds affecting the Colorado Plateau region blew chiefly towards the present south.

Outside of the Quaternary, there are no convincing cases of wind-blown silt in the geological record. It has been claimed, but appears very doubtful, that certain siltstones of the British Old Red Sandstone are wind-blown. The difficulties seem to lie with the criteria of recognition, rather than in the unlikelihood of finding loess in pre-Quaternary deposits.

READINGS FOR CHAPTER 3

Details of the relief and climate of major deserts can be found in any good geographical atlas. A useful summary of desert morphological features is to be found in:

SMITH, H. T. U. 1968. Geologic and geomorphic aspects of deserts. *Desert Biology*. Academic Press, New York and London, pp. 51–100.

STONE, R. O. 1967. 'A desert glossary.' *Earth-Sci. Rev.*, **3**, 211–268.

Our understanding of sediment transport by the wind still depends very heavily on the remarkable book:

BAGNOLD, R. A. 1954. *The Physics of Blown Sand and Desert Dunes*. Methuen and Co., London, 265 pp;

though the following papers are also of value:

CHEPIL, W. S. 1945. 'Dynamics of wind erosion.' *Soil Sci.*, **60**, 305–320, 397–411, 475–480.

SHARP, R. P. 1964. 'Wind-driven sand in Coachella Valley, California.' *Bull. geol. Soc. Am.*, **75**, 785–804.

WILLIAMS, G. 1964. 'Some aspects of the eolian saltation load.' *Sedimentology*, **3**, 257–287.

Many papers have been written on the textural properties of wind-blown sediments, of which the following are perhaps representative:

FRIEDMAN, G. M. 1961. 'Distinction between dune, beach and rivers sands from the textural characteristics.' *J. sedim. Petrol.*, **31**, 514–529.

HJULSTRÖM, F., SUNDBORG, Å., and FALK, Å. 1955. 'Problems concerning the deposits of windblown silt in Sweden.' *Geogr. Annlr*, **37**, 86–117.

MCKEE, E. D. and TIBBITTS, G. C. 1964. 'Primary structures of a seif dune and associated deposits in Libya.' *J. sedim. Petrol.*, **34**, 5–17.

SWINEFORD, A. and FRYE, J. C. 1951. 'Petrography of the Peoria Loess in Kansas.' *J. Geol.*, **59**, 306–322.

Representative accounts of wind accumulations, sand seas, and coastal dune complexes are to be found in:

AUFRÈRE, L. 1935. 'Essai sur les dunes du Sahara Algerien.' *Geogr. Annlr*, **17**, 481–498.

BAGNOLD, R. A. 1951. 'Sand formations in southern Arabia.' *Geogrl J.*, **117**, 78–85.

COOPER, W. S. 1958. 'Coastal sand dunes of Oregon and Washington.' *Mem. geol. Soc. Am.*, **72**, 169 pp.

HASTENRATH, S. L. 1967. 'The barchans of the Arequipa region, Southern Peru.' *Z. Geomorph.*, **11**, 300–331.

HOLM, D. A. 1960. 'Desert geomorphology in the Arabian Peninsula.' *Science*, **132**, 1369–1379.

INMAN, D. L., EWING, G. C. and CORLISS, J. B. 1966. 'Coastal sand dunes of Guerrero Negro, Baja California, Mexico.' *Bull. geol. Soc. Am.*, **77**, 787–802.

MABBUTT, J. A. 1968. 'Aeolian landforms in Central Australia.' *Australian Geographical Studies*, **6**, 139–150.
MADIGAN, C. T. 1936. 'The Australian sand-ridge deserts.' *Geogrl Rev.*, **26**, 205–227.
SHARP, R. P. 1963. 'Wind ripples.' *J. Geol.*, **71**, 617–636.
SHARP, R. P. 1966. 'Kelso Dunes, Mojave Desert, California.' *Bull. geol. Soc. Am.*, **77**, 1045–1074.

Data on the internal structures of wind-blown dunes are meagre, but the following accounts are especially useful:

LAND, L. S. 1964. 'Eolian cross-bedding in the beach dune environment, Sapelo Island, Georgia.' *J. Sedim. Petrol.*, **34**, 389–394.
McKEE, E. D. 1966. 'Dune structures.' *Sedimentology*. **7**, 3–69.
McKEE, E. D. and TIBBITTS, G. C. 1964. 'Primary structures of a seif dune and associated deposits in Libya.' *J. sedim. Petrol.*, **34**, 5–17.
SHARP, R. P. 1966. 'Kelso Dunes, Mojave Desert, California.' *Bull. geol. Soc. Am.*, **77**, 1045–1074.

Little work has been done on aeolian sediments older than the Quaternary. Amongst the most interesting accounts, however, are:

SHOTTON, F. W. 1937. 'The Lower Bunter Sandstones of North Worcestershire and East Shropshire. *Geol. Mag.*, **74**, 534–553.
SHOTTON, F. W. 1956. 'Some aspects of the New Red Desert.' *Lpool Manchr geol. J.*, **1**, 450–465.
STOKES, W. L. 1961. 'Fluvial and eolian sandstone bodies in Colorado Plateau.' *Geometry of Sandstone Bodies*, American Association of Petroleum Geologists, Tulsa, Oklahoma, pp. 151–178.

The first paper of Shotton's is also a classic in the literature of palaeocurrent studies.

CHAPTER 4

River Flow and Alluvium

4.1 General

Rivers feature significantly in the landscape of every continent but Antarctica. They are networks of channels, tree-like in plan, that collect and convey from the land the precipitation that falls upon it. The fingertip gullies and rills of river networks reach to the very heart of the largest continents, whilst the networks themselves, each occupying a definite drainage area or basin, in the case of the largest achieve sub-continental proportions. The Amazon system, for example, with a drainage area of $5{\cdot}7\times10^6$ km^2 and a discharge at the mouth of $1{\cdot}0\times10^7$ m^3/sec, drains approximately one-third of the total area of South America. The Mississippi, of comparable drainage area, serves a like proportion of North America, to cite another case.

Rivers also gather up and bear away the loose, solid debris and dissolved salts that result from the weathering crustal rocks. Sediment particles are moved down river channels by the force of the flowing water, itself maintained by gravity, and every year the rivers of the world discharge into lakes, seas and oceans a quantity of material estimated to total 2×10^{14} kgm. The annual discharges of the larger rivers are impressive. For example, that of the Hwang Ho is $1{\cdot}9\times10^{12}$ kgm, of the Amazon $0{\cdot}9\times10^{12}$ kgm, and of the Mississippi $0{\cdot}6\times10^{12}$ kgm. It would appear that a small number of comparatively large rivers are responsible for all but a minor proportion of the total sediment annually discharged into the seas and oceans.

However, the sediment borne off by rivers is not all destined for the sea bed. Although the upper reaches of river networks experience erosion, with bed rock occurring in the channel, the lower parts are commonly marked by a state of net deposition expressed as thick, laterally extensive spreads of alluvial sediment in which the active courses are embedded. These occurrences of alluvium include alluvial fans built against valley side and mountain fronts, the flood-plains of single streams in hill country, and coastal plains of alluviation formed where the flood-plains of separate river systems coalesce. The total land area underlain by the alluvium of modern rivers is far from negligible although comparatively small; some of the richest and most densely and longest populated lands of the world are alluvial. Individual spreads of alluvium achieve huge dimensions. Thus the alluvial valleys of the Mississippi, Ganges, Ob, Volga and Tigris-Euphrates have each an extent in the direction of stream flow of the order of 1000 km and a width of up to 200 km. One of the largest composite spreads of alluvium is that in the coastal plain of the north-western Gulf of Mexico. Between the Apalachicola River in the east and the Pánuco River in the west, Quaternary alluvium occurs in a belt 1600 km long and up to 130 km broad.

Alluvial landscapes have great variety, in spite of restrictions imposed by low relief, and can be read by an experienced eye as readily as grander and more startling scenery. The traces on the land of abandoned channels, and the distribution of basins holding swamps or shallow lakes, all tell of the construction of the landscape and of the character of subjacent sediments. Alluvial deposits are amongst the most varied of all the major sediment groups. They range from gravels and sands deposited from bed load in channels, to silts and clays carried in suspension and laid down on the open flood-plain, only to be sun-dried and colonized by plants and animals after the water subsided. Peat layers further complicate flood-plain sediments in areas that bear a dense and persistent plant cover. In hot dry climates, however, the alluvium contains beds of salts precipitated from the flood or ground water and may be deeply oxidized. The record of alluvial sediments goes far back beyond the Quaternary, for we find evidence of them in rocks dating from the earliest times.

4.2 River Networks and Hydraulic Geometry

There are quantifiable relationships between the geometrical and hydraulic properties of river channels and channel networks. These relationships are of interest in connection with stream processes and also in the interpretation of alluvium, for we find that evidence of channel shape and size is commonly preserved in alluvial landscapes and strata.

The shape of the drainage network, varying from trellis to dendritic, and the density of occurrence of channels in the network, depend on many factors, of which climate, bed rock, and geological structure are paramount. However, the length from the headwaters of a channel in the network is related to the total drainage area contributing to that measured length by the formula

$$L = 1{\cdot}2A^{0{\cdot}65}, \tag{4.1}$$

in which L is the channel length in kilometres and A is the drainage area in square kilometres. This equation with but little variation describes basins of very different sizes and locations. Since the exponent slightly exceeds 0·5, it follows that drainage basins elongate as they grow in size. Also of interest is that the mean annual discharge from a basin depends on basin area according to the equation

$$\bar{Q} = 0{\cdot}0086A \tag{4.2}$$

for $A < 5{\cdot}7 \times 10^6$ km^2, in which $\bar{Q}$ is the mean annual discharge in m^3/sec. The data forming the basis of this relationship show much scatter, however, according to climate.

In moving down a stream channel, the discharge of the channel increases and with it the channel width and depth. From investigations in many different basins, we can write for the downstream changes

$$w = k_1\,\bar{Q}^{0{\cdot}5}, \tag{4.3}$$

$$d = k_2\,\bar{Q}^{0{\cdot}4}, \tag{4.4}$$

and

$$U = k_3\,\bar{Q}^{0{\cdot}1}, \tag{4.5}$$

in which $\bar{Q}$ is the mean annual discharge at a station on the channel, w, d, and U are the width, mean depth and mean flow velocity, respectively, at that station, and k_1, k_2 and k_3 are constants for discharges equal to the mean annual discharge, depending on the basin. Since by continuity

$$\bar{Q} = wdU, \tag{4.6}$$

we find the exponents in Eqs. (4.3) to (4.5) sum to unity, and the constants give a product of unity.

Inspection of a channel network as portrayed on a map will show that it consists of channel segments of different degrees of importance as estimated using size. By custom we denote as first-order segments the smallest unbranched channels portrayed, though on account of the map scale these are not necessarily the smallest existing on the ground. The second-order segments are those formed by the joining together of first-order channels, and so on. We shall mean by n the order number of a channel segment, which is, of course, a positive integer. The basin containing the network can itself be categorized by assigning to it the order number of the highest-order segment it contains. We denote by m the order number of the basin, i.e. of the highest-order segment.

The study of many stream drainage basins has led to the following useful generalizations, sometimes called 'morphometric laws'

$$N_n = R_b^{(m-n)}, \tag{4.7}$$

$$L_n = R_L^n, \tag{4.8}$$

$$S_n = R_S^{(m-n)}, \tag{4.9}$$

and

$$A_n = R_A^n, \tag{4.10}$$

in which N_n, L_n, S_n, and A_n are the number, mean length, mean slope, and mean drainage area, respectively, of segments of order n. The symbols R_b, R_L, R_S, and R_A, denote, respectively, the bifurcation ratio, stream length ratio, stream slope ratio and drainage area ratio, which are defined by

$$\left.\begin{aligned} R_b &= \frac{N_n}{N_{(n+1)}}, & R_L &= \frac{L_{(n+1)}}{L_n} \\ R_S &= \frac{S_n}{S_{(n+1)}}, & R_A &= \frac{A_{(n+1)}}{A_n} \end{aligned}\right\}. \tag{4.11}$$

These relationships, which are all geometric series, allow many interesting calculations to be done on drainage networks and basins. But Eqs. (4.7) and (4.8) at least are without doubt partly abstractions, for they would describe any topologically random branching network, i.e. of roots or branches of a tree or of blood vessels. However, the values of R_b and R_L appear to be geomorphologically significant and to depend on the degree of dissection of the terrain. For example, R_b varies between about 2·5 and 7·0 for natural basins but is largest for deeply dissected mountainous ones.

4.3 Channel Flow and Sediment Transport

As observed at a single station, the quantities of water and sediment discharged daily by a river vary with the season to an extent dependent on the general climate of the river basin. A basin sited in a hot, dry climate may have a zero flow for many days on end, but a very large flow for a few hours or days after a storm. In regions of high rainfall and humidity, however, the daily discharge varies little from one day or season to another.

We customarily describe the aqueous flow measured at each station by means of a flow-duration curve, which is simply a discharge frequency curve compiled from values of the daily mean discharge (Fig 4.1). Such a curve shows three especially significant discharge values. The median discharge is that discharge equalled or exceeded fifty per cent of the time, i.e. half the observation days have a discharge exceeding the median. Larger still is the mean annual discharge, already referred to, which is the arithmetic mean of all individual daily mean discharges in a year of measurement. The bank-full discharge is the largest of the three and is that discharge

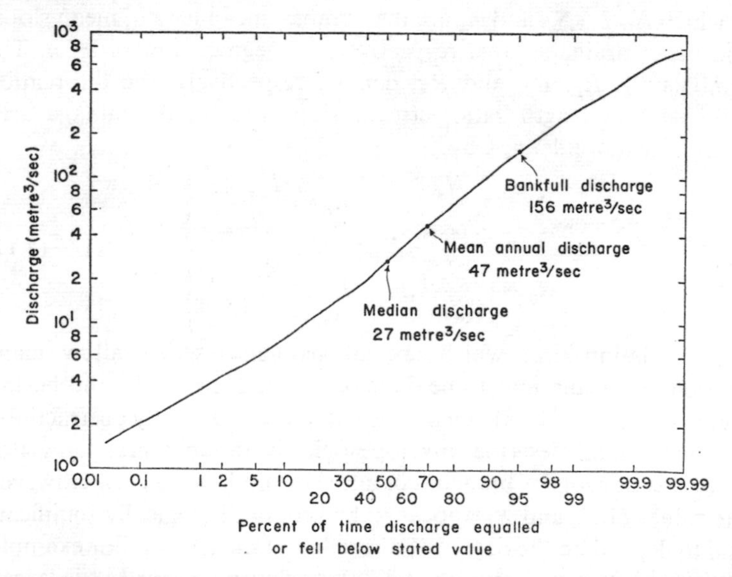

Fig 4.1 Representative flow duration curve of a river plotted on logarithmic probability paper.

which, if exceeded, would cause the river to spill over its banks on to the flood-plain. All discharges larger than bank-full are flood discharges for the station. A flood discharge can be specified by its recurrence interval, which is larger the more severe the flood. The bank-full discharge has a recurrence interval of 1·56 years on average with a range between 1 and 5 years. The recurrence interval in years of a flood discharge is given approximately by $(Q_f/Q_b)^{2\cdot 8}$ where Q_f is the discharge at the required recurrence interval and Q_b is the bank-full discharge.

Since discharge at a station is a function of time, the width, mean depth and mean velocity of the river must also vary with time. Their variations with instantaneous discharge follow power laws differing from Eqs. (4.3) to (4.5) only in terms of the values of the constants and exponents involved. For a large sample of river gauging stations biased towards semi-arid conditions, it was found that width, mean depth, and mean flow velocity increased as the 0·26, 0·40 and 0·34 powers, respectively, of the discharge. Thus the flow of a natural river is unsteady, for both the mean velocity and depth at a place are functions of time. One has but to walk along the bank of a river to see that the flow is also non-uniform, i.e. at a given time, flow depth and mean velocity vary in the current direction. If depth decreases but velocity increases down a reach, the flow is classified as accelerated. The flow is said to be decelerated, however, if depth increases but velocity decreases in the flow direction. The flow of natural rivers, pre-eminently under seasonal control, is therefore varied and unsteady. The variation is gradual within a reach as a whole, but rapid where the channel suddenly deepens or shoals. The unsteadiness arises from changes of discharge, at certain times rapid but at others gradual, as flood waves build up and pass downstream. The variedness depends on channel shape and bed configurations of various types.

A simple mathematical model of river flow can be made only if we assume a steady, uniform flow. There will then be no mean accelerations in the system, and the gravity force, which propels the flow along, will be exactly balanced by friction between the moving water and the fixed channel boundaries.

Consider an aqueous stream of uniform depth d in a rectangular channel of uniform width w and declination β (Fig 4.2). The downslope component of the gravitational force acting on the water in a channel segment of length L is

$$F_g = \rho g\, dwLS, \tag{4.12}$$

in which ρ is the density of water, g is the acceleration due to gravity, and S is the slope written instead of $\sin\beta$ as β is small. Now the frictional resisting force F_D, equal and opposite to F_g, is

$$F_D = \tau_0[L(2d+w)], \tag{4.13}$$

in which τ_0 is the tangential resisting stress, or boundary shear stress, per unit area of the wetted boundary. Equating (4.12) and (4.13), and putting $dw = a$ (the cross-sectional area of the flow), we write

$$\tau_0 = \rho g S \frac{a}{(2d+w)}, \tag{4.14}$$

in which $a/(2d+w)$ is defined as the hydraulic radius R. The hydraulic radius is simply the cross-sectional area of the flow divided by the wetted perimeter, and in the case of a broad shallow channel is nearly equal to flow depth.

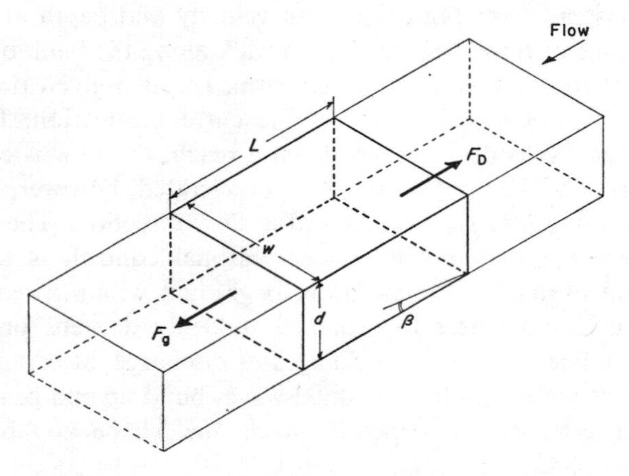

Fig 4.2 Definition diagram for flow in an ideal river channel.

Now, from dimensional reasoning first applied to flow in pipes, we can write for channel flow

$$\tau_0 = \frac{f}{4}\rho\frac{U^2}{2}, \tag{4.15}$$

where f is the Darcy-Weisbach resistance coefficient for pipe flow and U is the mean velocity in the channel. Combining Eq. (4.14) with Eq. (4.15), we have that

$$f = \frac{8gSR}{U^2}, \tag{4.16}$$

and

$$U = \sqrt{\frac{8g}{f}}\sqrt{(SR)}. \tag{4.17}$$

Eq. (4.17) commonly appears in the form

$$U = C\sqrt{(SR)}, \tag{4.18}$$

in which the Chézy resistance coefficient C is not dimensionless.

Experimental investigations confirm the conclusions of dimensional reasoning that f is a function of Reynolds number and boundary roughness, and afford extensive data on how f varies. In laminar flow, f decreases rapidly with ascending Reynolds number. The decrease of f with Reynolds number is, however, rather gradual in the turbulent regime. Of greater significance in turbulent flow is the rapid increase of f with ascending roughness of the boundary, as measured by the ratio of the size of the roughness elements (irregularities of surface, i.e. sediment particles or bed forms) to the flow depth.

In open-channel flow turbulence sets in at a Reynolds number in the approximate range $500 < Re < 2000$. Inspection of Fig 4.3 will make it clear that turbulent flow is the only type of flow of practical interest in connection with rivers. But since a river has a free surface, we must also take gravitational forces into account, by reference to the Froude number (see Ch. 1.2). The flow in an open channel is described as *super-critical* when $Fr > 1$, but *sub-critical* when $Fr < 1$. The graph of $Fr = 1$ is also plotted in Fig 4.3, whence there are four regimes of open-channel flow: sub-critical laminar, super-critical laminar, sub-critical turbulent, and super-critical turbulent. Only the last two are of practical interest, and of these, the last-mentioned is the least common.

The local mean fluid velocity in a river channel increases with growing distance from the bed and walls. If the channel is much broader than deep, the velocity profile is given approximately by von Kármán's universal law (Eq. 1.25) except in that part of the flow close below the free surface. The velocity maximum which appears in the profile occurs just below the free surface in a broad

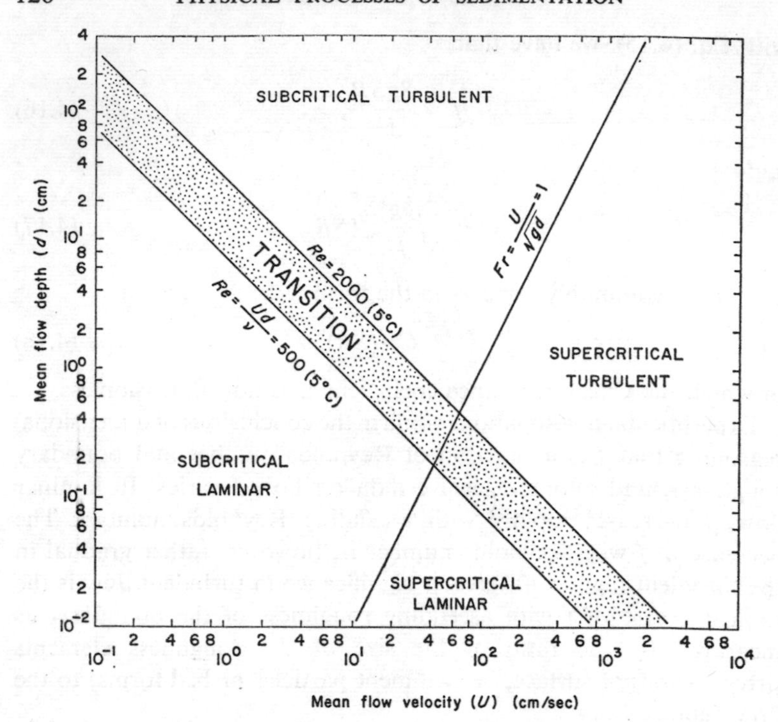

Fig 4.3 Regimes of flow in open channels.

shallow channel. When channel depth and width are of the same order, however, the position of the velocity maximum is shifted downward relatively.

We saw in Ch. 1 that a fluid flow could be regarded as a transporting machine with respect to granular solids borne along in the flow. We also saw that a sediment transport rate, multiplied by a notional conversion factor, is a rate of doing work that can be related through the efficiency to the available power, just as in the case of any man-made working machine. The sediment transport work rates for steady river flow can be written

$$\text{bed load work rate} = \frac{(\sigma-\rho)}{\sigma} g m_b U_b \tan\alpha = i_b \tan\alpha, \qquad (4.19)$$

$$\text{suspended load work rate} = \frac{(\sigma-\rho)}{\sigma} g m_s U_s \frac{V_0}{U_s} = i_s \frac{V_0}{U_s}, \qquad (4.20)$$

in which σ is the density of the solids, ρ is the fluid density, m_b is the mass of bed load above unit area of the bed and U_b is its mean transport velocity, m_s is the mass of suspended load above unit area of the bed and U_s is its mean transport velocity, and V_0 is the free falling velocity of the suspended solids. The notional conversion factor for bed load is $\tan\alpha$, the ratio of tangential to normal components of the grain momentum arising from collisions between grains. The factor for the suspended load is V_0/U_s. Now in the case of a river the power available for sediment transport is easily specified as

$$\omega = \tau_0 U = \rho g S R U = \tfrac{1}{8} f \rho U^3, \tag{4.21}$$

where ω is the mean available power supply to the column of fluid above unit bed area. Introducing dimensionless efficiency factors e_b for bed load and e_s for suspended load, we obtain for the total load transport rate

$$i = i_b + i_s = \omega\left[\frac{e_b}{\tan\alpha} + \frac{e_s U_s}{V_0}(1-e_b)\right], \tag{4.22}$$

whence from Eqs. (4.15) and (4.21), the transport rate is proportional to the cube of the mean flow velocity or to the 3/2 power of the boundary shear stress.

The above transport function applies provided the dimensionless bed shear stress θ exceeds 0·4, and is in practice remarkably accurate as such functions go. The value of the bed load efficiency factor is about 0·13, and is slightly influenced by the mean flow velocity and calibre of load. The suspended load efficiency factor is much smaller, about 0·015.

The sediment load borne by a river varies vertically as regards concentration and calibre. The bed load, confined to the lowest parts of the flow, ordinarily consists of densely arrayed sand grains and sometimes small pebbles. Cobbles and boulders are moved only under extreme conditions of very large boundary shear. Within the bed load zone, the sediment concentration decreases upwards but is everywhere relatively high. Grain size decreases upwards in the case of a load of fine sand grains that shear viscously. Coarser loads display a partially or fully inertial shearing, however, and in their case grain size increases upwards within the bed load zone. The concentration of suspended sediment is ordinarily low and, for particles of a single free falling velocity V_0, varies as

$$\frac{C_y}{C_a} = \left(\frac{d-y}{y} \cdot \frac{a}{d-a}\right)^z \tag{4.23a}$$

$$z = \frac{V_0}{\kappa k_4 \sqrt{(gRS)}}, \tag{4.23b}$$

in which C_y is the concentration at a distance y from the bed, C_a is the concentration at a reference distance a, d is the flow depth, κ is the von Kármán constant, and k_4 is the ratio of the sediment and momentum diffusion coefficients. The equation, following from the simple diffusion analysis sketched in Ch. 1, is kinematic in basis and states that the concentration is infinitely large at the bed but decreases upwards in the flow. Neither assertion is in complete accord with experience of suspended loads of more than one size of grain, for the concentration at the bed is never infinitely large, and it is commonly found, especially for flashy streams, that the concentration of the smaller particles actually increases upwards. If, however, the volume occupied by sediment particles is taken into account, we can obtain from the basic diffusion equations a statement permitting for a mixed load an *upward* increase in concentration for certain fines and a bed concentration of the order of unity. In a suspended load consisting of mixed particle sizes, the fines are, as it were, squeezed into the upper parts of the flow by the bulky particles of large V_0 concentrated near to the bed. In this way the concentration of the smaller particles in suspension increases upwards whilst that of the larger ones decreases upwards.

The calibre of the stream bed load, carefully averaged across the line of flow, also decreases in the downstream direction, according to the equation

$$D = D_0 e^{-k_5 L}, \tag{4.24}$$

in which D is the mean particle diameter at a distance L along the channel, D_0 is the mean particle diameter where $L = 0$, e is the base of natural logarithms, and k_5 is a constant related to the downstream change of stream power.

4.4 Channel Form and Process

Channel pattern is the plan view of a reach of a river, and includes relatively straight, braided and meandering channels. Braided and meandering patterns are by far the commonest assumed by rivers flowing in their own alluvium. Relatively straight channels are rare, and can usually be explained in terms of control by local geological structures. A meandering river has a single channel that repeatedly swings first to one side then the other of its mean flow direction. The degree of meandering, or oscillation, is measured by the sinuosity, the ratio of channel or talweg length to the length of valley occupied by the channel. For meandering streams, the sinuosity is greater than 1·5. A braided stream is one whose channel shows successive diversion and rejoining of segments of the channel around alluvial islands. But there is no clear cut division between meandering and braided streams, as witness streams whose single, winding channel is divided from place to place by a mid-channel island. The most representative of braided rivers are perhaps those found in glacial outwash plains; their channels are divided across the line of flow into many curved segments by numerous alluvial islands. The sinuosity of channel segments of braided streams is less than 1·5.

It is far from clear why some streams meander whilst others are braided. The answer cannot be a question of size, as braiding and meandering are known from the smallest as well as the largest rivers. Very probably the solution lies in the way in which each particular river adjusts to the externally controlled discharge and amount and calibre of the sediment load. This adjustment involves the flow resistance of the channel, which is made up of varying contributions from the particles in the bed and banks, the shape and size of bed forms and minor channel irregularities, and the large scale channel configuration. Empirically, however, we can distinguish meandering from braided channels in terms of slope and discharge, according to the equation

$$S = 0{\cdot}013\, Q_b^{-0{\cdot}44}, \qquad (4.25)$$

in which Q_b is the bank-full discharge in cubic metres per second. For a given discharge, braided streams occur on slopes steeper than the slope given by Eq. (4.25), whilst meandering streams occur on gentler ones. Also, braided streams generally have coarser material in the bed and banks are relatively shallower and broader than meandering ones.

The oscillatory curving of river channels, whether braided or meandering, can probably be attributed to the transverse instability of a channel flow confined within deformable boundaries at the high

Reynolds numbers typical of natural streams. This general problem has so far been attacked analytically and empirically only within the context of meandering streams, whose channel geometry is relatively simple. One analytical solution, based on stability theory, gives for the meander wavelength λ the equation

$$\lambda = k_6 Q^{0\cdot525} D^{-0\cdot316}, \tag{4.26}$$

in which k_6 is a constant, Q is a measure of discharge, and D is the mean diameter of the bed and bank material. Other analyses depend on the downstream convection of an assumed transverse, oscillatory disturbance of the channel flow, and give equations of the form

$$\lambda = k_7 \sqrt{a}, \tag{4.27}$$

in which k_7 is a constant partly controlled by channel resistance, and a is the cross-sectional area of the flow. Now since $a = Q/U$ and U is almost independent of Q (see, for example, Eq. 4.5), we have that

$$\lambda = k_8 Q^p, \tag{4.28}$$

where k_8 is a new constant and the exponent $p \approx 0\cdot5$, the precise value depending on how U and Q are related. It has, of course, been empirically recognized for many years that meander wavelength is proportional to discharge raised to a power close to 0·5.

There are, however, problems in objectively defining meander wavelength for natural channels and in choosing the most significant measure of discharge. Recently, it was found that some natural streams display a spectrum of meander wavelengths of which as many as three could be described as dominant. We also know that meander wavelength correlates better with some measures of discharge than others, and differently with some measures than others. We find empirically that, with λ in metres and the discharge in cubic metres per second,

$$\lambda = 168 \bar{Q}^{0\cdot46}, \tag{4.29}$$

$$\lambda = 126 \bar{Q}_{\mathrm{mm}}^{0\cdot46}, \tag{4.30}$$

and

$$\lambda = 22\cdot6 Q_{\mathrm{b}}^{0\cdot62}, \tag{4.31}$$

in which $\bar{Q}$ is the mean annual discharge, $\bar{Q}_{\mathrm{mm}}$ is the mean maximum monthly discharge, and Q_{b} is the bank-full discharge. The variance of the data proved to be least using mean annual discharge and greatest using bank-full discharge. It seems, therefore, that discharges close to the mean annual discharge in value are most effective in shaping channel bends.

The flow of water in a channel bend gives rise to an excess of fluid pressure on the outer bank and a corresponding deficit over the inner. Throughout the main body of the flow the radial pressure gradient thus set up is balanced by the gradient of the centrifugal force. At the bed, however, the radial pressure gradient is unbalanced on account of frictional losses in the flow, and so water moves inwards over the bed towards the inner bank, down the radial pressure gradient. Continuity demands that its place be taken by a

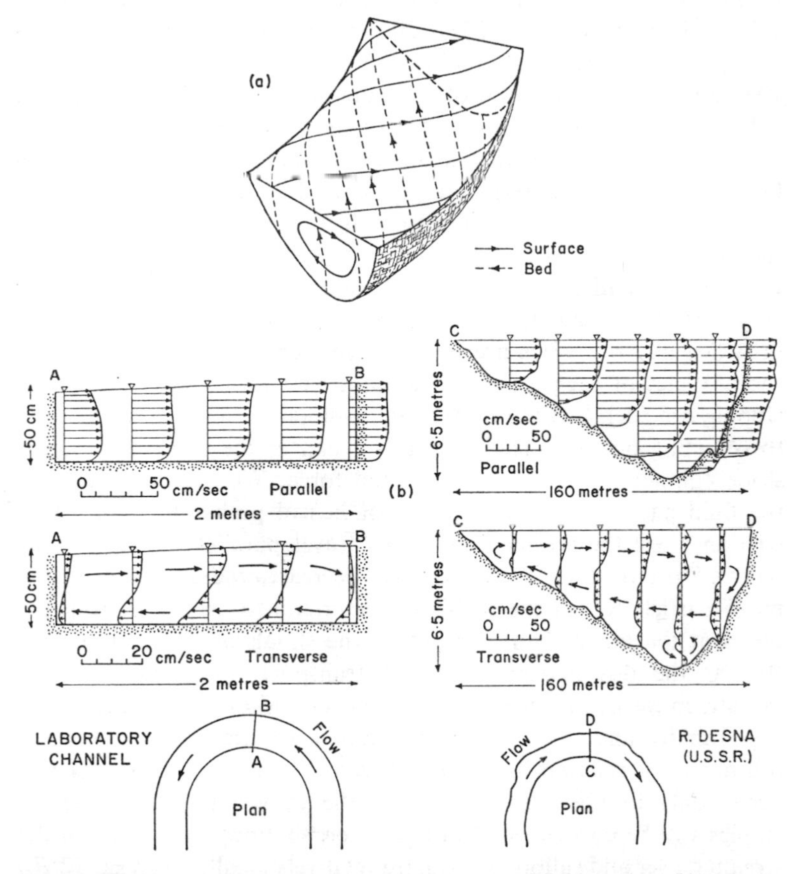

Fig 4.4 Fluid motion in channel bends. (a) Idealized fluid motion. (b) Velocity profiles parallel and transverse to the mean flow observed in a laboratory channel and a medium-sized river (data of I. C. Rozovskii).

compensatory flow from the inner to the outer bank over the upper part of the flow cross-section. In a channel bend, therefore, a water particle follows a helical path which carries it from surface to bed and from inner bank to outer (Fig 4.4a). Furthermore, the velocity of the particle can be resolved into components parallel and transverse to the channel centre-line. Fig 4.4b gives examples of velocity profiles measured in experimental and natural channels. It will be noted that the pattern of velocities is qualitatively similar in the two cases, in spite of the fact that the experimental channel is rectangular while the natural one is triangular, being an adjustment to imposed duties. It will also be seen that the maximum transverse velocity is of the order of 10 or 20 per cent of the maximum downstream component. The 'pitch' of the helical path is therefore about 10 or 20 times the channel width, i.e. about one meander wave length.

Further inspection of Fig 4.4b will show that, in a channel bend, the local velocity gradient and mean velocity, and hence the mean boundary shear stress and stream power, all decrease from the outer to the inner bank. In an equilibrium channel, therefore, we may plot the skin-friction lines that form a part of the helical flow, together with lines parallel to the channel banks of equal stream power (Fig 4.5). Assuming that the channel discharges bed-load material in which all grain sizes are equally represented, we shall find that the particles of each grain size assume that radial position on the sloping bed such that the transverse up-slope component of the fluid force acting on the particles exactly balances the transverse downslope component of the gravitational force. We see that although the fluid particles follow portions of helical paths along the skin-friction lines, the sediment particles travel parallel to the channel banks. Since the local fluid bed force decreases radially inwards, the mean particle size of the bed-load material also decreases from the deeps at the outer channel bank to the shoals at the inner curve. But as the downstream sediment transport rate also decreases radially inwards, the mean size of the bed-load material decreases at first rapidly and then more gradually over the shoaling bed. It follows from relationships summarized in Fig 2.6 that dunes will occur only in the deeper parts of the curved channel, whereas ripples will be concentrated in the inner, shallower parts where the stream power and calibre of load are relatively small. From Eq. (2.20) we might expect the height of dunes to decrease radially inwards.

In practice, channel bends suffer continual downstream and lateral movement and so are never in strict equilibrium. Erosion takes place at the steep outer bank, where the flow exerts large shear stresses, while a compensatory deposition occurs on the inner bank where shear stresses are smaller and the bottom flow is directed inwards. The rate of erosion, measured as the rate of bank recession, depends on the nature of the bank materials and the power of the stream. Recession is most rapid in the case of banks of loose sand or gravel and streams of large power, and is least rapid in the case of banks of silt or clay and low-powered streams. Actual rates of bank recession vary from a few decimetres to many tens of metres

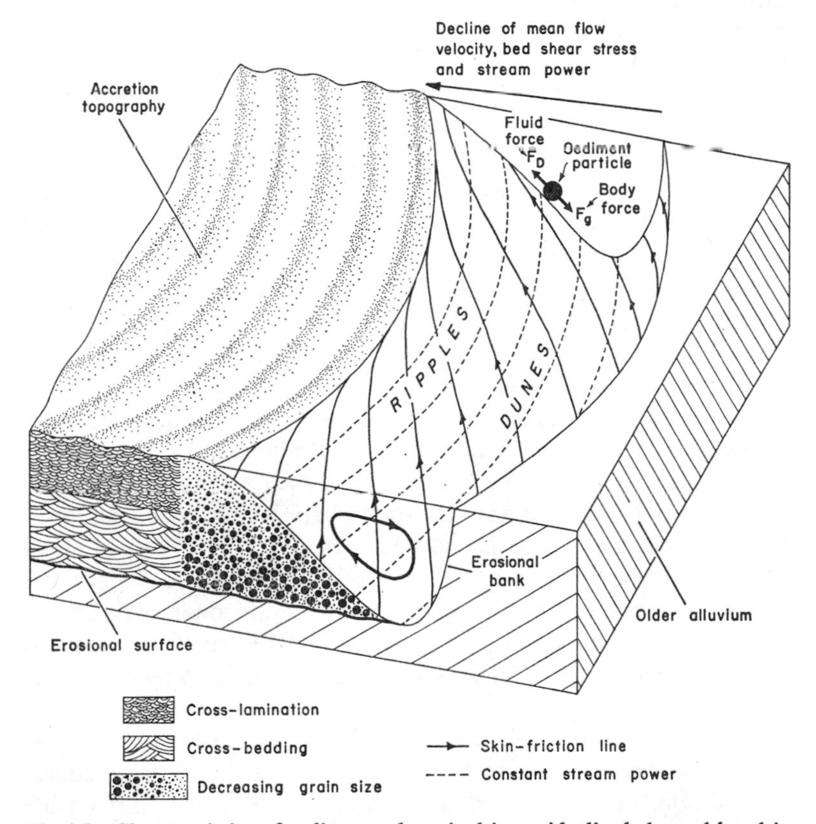

Fig 4.5 Characteristics of sediments deposited in an idealized channel bend in relation to pattern of fluid motion and the chief hydraulic quantities. For equilibrium of the sediment particle shown in the cross-section, the upslope fluid force F_D must be equal and opposite to the downslope body force F_g.

per year. The recession generally takes place in steps, as more or less large slices of bank slump into the channel.

The shifting of channel segments in braided streams is rarely marked by any distinctive sediment bars, other than the alluvial islands, or braid bars, between the segments. In the case of meandering streams, however, channel shifting is revealed in a pattern of curvilinear scroll bars which together form the point bar enclosed by each meander loop. Scroll bars vary in transverse spacing from a few metres to a few hundred metres and in elevation above the adjacent swales from a few decimetres to 2 or 3 metres. A scroll bar begins to form against the inner bank at the upstream end of the meander loop. In the course of time the bar is built upwards and extended further downstream on a line parallel to the inner bank but separated from it by a narrow swale of shallow depth. Very commonly, a second scroll bar commences to form at the upstream end of the loop before the growing tip of the preceding bar has reached the downstream end. On occasions, three or four simultaneously growing scroll bars can be found at the edge of one point bar. Thus a scroll bar does not generally represent an annual increment of sediment on the point bar, though it undoubtedly grows most rapidly during and just after floods.

Rivers change their courses in more dramatic ways than by bank recession. In braided streams, new channel segments and new braid bars arise as runnels crossing alluvial islands are enlarged. Existing channels become blocked off at one end, and so backwaters are formed. Channel shifting occurs in two ways in meandering streams. In neck cut-off, a meander loop is abandoned when a new channel is cut across the spit of land at the root of a point bar. The abandoned loop is plugged at the ends and becomes an ox-bow lake. In chute cut-offs, however, only a part of the loop is abandoned, on account of the enlargement into a channel of a swale on the point bar surface. But the most dramatic changes occur by avulsion, which is the sudden shifting of a whole stream course, built up by deposition to a disadvantageously low gradient, to a new site in an adjacent and more favourable area. Rivers as diverse as the Mississippi, Rio Grande, Indus and Hwang Ho have all repeatedly changed course by avulsion, often with a disastrous attendant loss of human life and property.

Concealed by the water in river channels are a variety of bed forms of a relatively smaller scale than braid or point bars. The larger of these additional forms, with dimensions of the order of the width and depth of the containing channel, are transverse, linguoid and barkhan-shaped bars. Transverse and linguoid bars are chiefly found in sand-bed streams. They have gentle upstream slopes but steep downstream faces at the angle of rest of the bed-load material. Transverse bars are regularly spaced along the channel and their crests, in plan varying from obliquely rectilinear to convex downstream, extend across the flow for almost the full channel width. Linguoid bars, whose crests also are convex downstream, are found in *en echelon* array in relatively broad, shallow channels. The barkhan-shaped bars appear to be restricted to gravel-bed streams, about which comparatively little is thus far known. In plan these bars resemble the barkhan dunes of the desert but are rather more compressed laterally. The downstream face is inclined apparently at the angle of rest of the gravel and in height is between 10 and 20 per cent of the flow depth.

The smaller of the generally concealed bed forms are current ripples, dunes, plane beds and antidunes. They are mainly restricted to sand-bed streams. An important contribution to the total resistance of river channels comes from these bed forms. When the bed form is ripples or dunes, the resistance arising from the bed forms is relatively large and the flow is said to be in the lower regime of channel roughness. If antidunes or plane beds are the bed form, the resistance attributable to the bed forms is relatively small and the flow is assigned to the upper regime of roughness. The existence of two regimes of roughness dependent on bed form is important, for when channel discharge and width are constant, a flow in the upper regime will be shallower than a flow in the lower regime of bed roughness. It is important to stress that the change from lower to upper regime takes place at a Froude number substantially less than $Fr = 1$.

Regarding ripples and dunes, we saw in Ch. 2 (Fig 2.6) that ripples form at relatively low stream powers and are restricted to very fine to medium sands, whereas dunes arise at moderate values of stream power and are not restricted as to sand size. The ripples found in river channels are chiefly of the linguoid form, bow-shaped varieties being especially common. Straight or sinuous crested ripples occur only in protected areas or in sluggish flows, for which the stream power is not much above the critical. Dunes are a common bed form in sand-bed streams. In the Mississippi, Niger and Senegal, for example, they are continuously recorded over hundreds of kilometres of river bed. On average, the height of river dunes is between 10 and 20 per cent of the mean flow depth, while the wavelength is several times the mean depth. In a big river, dunes between 2 and 5 metres in height and 50 and 200 metres in wavelength are likely to be common. The dunes are mostly three-dimensional. They have sinuous or short and strongly curved crests, and their troughs contain deep hollows scoured out by complex separated flows to lee. But as a massive redistribution of sediment is called for if a dune bed is to be substantially modified in appearance, we find that river dunes are never fully adjusted to the prevailing flow conditions, which as we saw are not constant in time. There is always a temporal lag between a change of flow and a corresponding change of bed form. This lag exists for all bed forms generated by an unsteady flow, but is particularly important for dunes. Thus it is not uncommon to find that dunes continue to grow in height and wavelength long after the peak of a flood has passed the site of the dunes.

Plane beds are the dominant bed form in few rivers with sand bottoms. Most of those recorded belong to the upper phase, of relatively large stream powers, shown in Fig 2.6 and to the upper flow regime.

Antidunes are commonest in broad, shallow channels of relatively steep slope carrying sand or gravel. As we have seen in Ch. 2, they are low-amplitude bed waves broadly in phase with similar waves on the water surface, and when they arise the Froude number is in the neighbourhood of unity. Antidunes form at stream powers in the range appropriate to the upper phase of plane beds and higher. They denote a large sediment transport rate, and are more characteristic of braided than meandering streams.

4.5 Flood-Plain Form and Process

A flood-plain is a strip of alluvial land that borders a stream channel and is periodically inundated by flood waters emanating from the channel. The flood-plain of a well-developed braided stream is not a continuous region, for it consists of the many alluvial islands or braid bars which divide up the flow. Moreover, the elements of the flood-plain have little permanency, for the sediment bars experience a continual and rapid modification by the flow passing round them.

Of greater importance are the often highly diversified flood-plains of meandering streams, but we must go beyond purely physical processes in order to understand their character and origin.

The active channel of a meandering stream is generally found to surmount an alluvial ridge that rises several decimetres or metres above the adjacent lowlands known as flood-basins. Within the alluvial ridge are active and abandoned meander loops enclosing point bars whose surfaces display the ridge and swale topography associated with scroll bars. It is also noticed that the active as well as the abandoned portions of the river channel are bordered over considerable distances by ridges of wedge-like cross-section called levées. The levées slope up gradually from the flood-basins surrounding the alluvial ridge, and culminate at the bank of the active or abandoned channel that they border. Levées vary in width between one-half and four times the channel width and in elevation range between a few decimetres and as much as 8 metres, depending on river size and calibre of load. The downslope gradient of levées is a few to many times the slope of the adjacent channel. Here and there the levées are cut transversely by sizeable channels, called crevasses, which serve to route flood water from the main river channel on to the flood-plain. At the levée crest the crevasse is a single, deep and wide channel. Down-slope it parts into smaller distributaries which surmount a slightly elevated fan or lobe-shaped mound of sediment called a crevasse splay. Very commonly, these splays extend well beyond the levée toe into the flood-basin.

Flood-basins are the lowest-lying parts of flood-plains. Their length is much greater than their width, and they are elongated in the direction of stream flow in the alluvial ridge. Often flood-basins are segmented into subsidiary basins connected across sills formed by large crevasse splays or by older alluvial ridges abandoned after avulsion. Flood-basins possess internal drainage systems composed of small channels. In arid and semi-arid regions, flood-basins are dry and exposed to wind action except during and shortly after floods. The plant cover of the basin is sparse or lacking, though a dense cover may be found on the bordering levées. Where the climate is wet and humid, however, river flood-basins include large vegetated marshes or swamps and even shallow, permanent lakes and ponds. In such regions the plant cover is dense, permanent and extensive. In their subtly changing character, the plant communities reflect the varying nature of the flood-plain environments, from the well drained levées to the poorly drained flood-basins. The plant cover promotes deposition from flood waters, by checking the flow, and makes its own contribution to the flood-plain sediments through the death and replacement of individuals.

The sedimentary regime of river flood-plains has as its most basic feature the repeated submergence and emergence of the sedimentary surface. A maximum of four stages can be recognized in this cycle, as follows: (i) the spilling of flood water from the main channel into empty flood-basins, (ii) the filling up of flood-basins to a stage where sustained flow down the flood-plain is possible, (iii) the emptying of flood-basins, and (iv) the drying out of flood-basins and modification of the newly deposited sediment.

The waters of a rising river gain the flood-plain earliest by way of unsealed crevasse channels and, at a slightly later stage, by spilling over uncrevassed stretches of the levées. In either case, the flow down the levées is fast and destructive, for the levées slope much more steeply than the channel from which the flood emanates. The flow in the crevasse channels themselves is commonly super-critical. Sheet erosion, channel widening and deepening, and widespread destruction of any plant cover are the normal consequences of the first phase of the flood-plain cycle.

Entry on the second phase is marked by the filling up of subsidiary flood-basins and their interconnection by submergence of any sills until the flood waters obtain free passage down the full length of the flood-plain. This phase is the most important as regards flood-plain deposition. Unless a second flood wave passes down the channel, erosion becomes restricted to a few areas close to the main channel. An insight into the processes operating in this phase of the flood-plain cycle is afforded by model studies of the flow in a rectangular channel bordered by a flat flood-plain (Fig 4.6). The mean velocity measured over a vertical is greatest at the centre of the submerged channel and decreases with ascending distance from the channel centre-line. Above each margin of the submerged channel, a line of large vortices with vertical axes is observed. These vortices are spaced out along alternate sides of the channel and their average down-stream separation distance is about twice the channel width. The vortices, representing a Kelvin-Helmholtz type of instability, serve to transfer high-velocity fluid from the main channel to the shallow flood-plain, where its momentum is largely destroyed. Thus the power of the flow over the flood-plain decreases from the channel margins outwards. Applying these findings to a natural channel carrying sediment, we would expect the rate of deposition on the flood-plain to decrease with increasing distance from the channel edges. Also, we would expect the mean size of the deposited sediment to fall in the same direction. Moreover, because we are dealing with an over-bank flow, we would expect the deposited sediment to be that carried in suspension by the stream, i.e. silt and clay but only the finest sand. The positioning of active channels within alluvial ridges is consistent with the above reasoning. We shall find below that flood-plain sediments are mostly fine grained.

The decay of the flood wave marks the third phase of the flood-plain cycle. As the flood subsides, the velocity of the flow over the flood-plain declines, until the sills re-emerge and ponding ensues. It is now that the finest suspended sediment settles out. The wind playing on the surface of the ponded flood waters generates waves that shape sandy bottom sediments into symmetrical ripple marks and cut low cliffs into the levée and crevasse-splay slopes.

The retreat of the flood brings to light a landscape of sand splays and mud flats on which plants and animals may readily encroach.

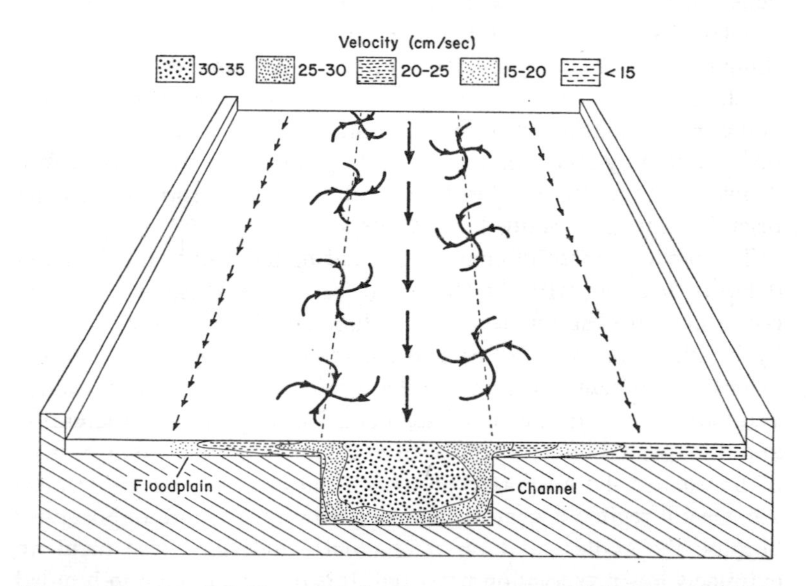

Fig 4.6 Pattern of fluid motion (surface vortices emphasized) and velocity distribution in a laboratory river flowing down its channel and over the surrounding flood-plain (data of J. Sellin). The flood-plain is approximately 45 cm wide.

The muddier sediments can, while soft, receive the imprints of rain drops and animals' feet, though they rapidly harden, shrink and crack under the drying sun and wind. The wind, too, sifts and winnows the sand and smaller mud chips, sometimes building up tall dunes at the edges of the sediment-covered areas. In wet, humid climates, the drying out of the flood-plain is never completed, but in dry climates the process commonly goes to the stage when salts are precipitated in the alluvial soil.

4.6 Deposits of Braided and Meandering Streams

The deposits of braided streams are relatively simple in character (Fig 4.7a). Texturally, they comprise varying proportions of gravel, sandy gravel, pebbly sand and the coarser grades of sand. Silt or clay beds are rare and in any case thin. The sands and gravels occur in beds between a decimetre or two and a few metres in greatest thickness. The beds are chiefly lens-shaped channel fills, and are seldom of uniform lateral thickness. Some gravel and sandy gravel beds are cross-bedded, whereas others show no internal bedding, though a well developed particle fabric, that involves upcurrent shingling, is commonly evident. Either even lamination or cross-bedding is found in the sand and pebbly sand beds. Ripple marks and cross-lamination are rare or absent. The infrequent silt and clay beds commonly show sun cracks, and less often the imprints of rain drops and animals' feet. Drifted wood and plant debris is the chief organic element in braided stream deposits.

The predominance of coarse bed load material in braided stream deposits, together with the structural features of these deposits, are consistent with the relatively steep slope and large power exhibited by braided streams. The beds preserved have properties denoting extensive sediment reworking, and are chiefly the remnants of islands and bars generated as channel segments formed, wandered laterally, and disappeared under a regime of rapid and continual change. Beds of fine material transported in suspension are rare, partly because the general regime is against their deposition, but more importantly because the endless and rapid channel wandering gives them an extremely low preservation potential. It is not uncommon in braided stream deposits to find that gravel-sized mud clasts are the only witness to the deposition of beds of silt or clay. Beds of fine sediment deposited in cut-off channels have, however, a greater preservation potential than those laid down on the tops of bars.

The deposits of meandering rivers are far more varied (Figs 4.5, 4.7b). The sediments formed in the channel, chiefly by the deposition of bed load materials on advancing dunes or ripples, are preserved in scroll bars aggregated into point bars. Sand is predominant, and gravel, silt and clay rare. The gravel is mainly found immediately above the surface of erosion swept out by the meandering channel during its lateral movement, as either a pebble pavement or a series of lens-shaped beds; large clasts of silt or clay-sized material are not uncommon in these basal gravels. The sands, which grow finer

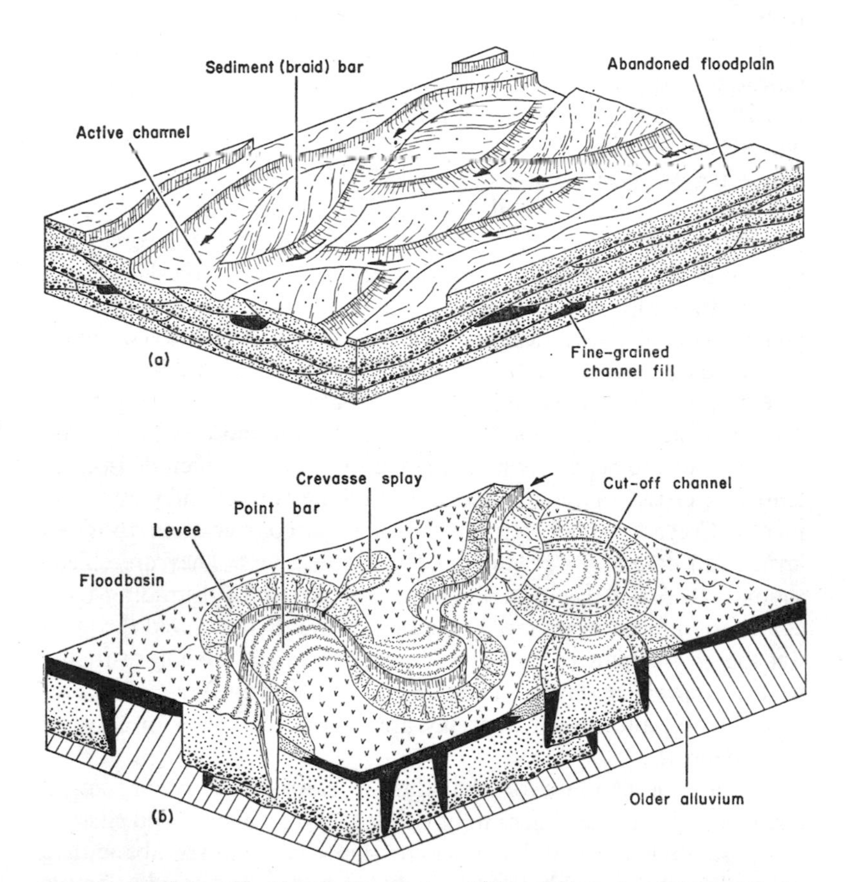

Fig 4.7 Models of fluviatile sedimentation. (a) Braided stream. (b) Meandering stream. Vertical scales exaggerated.

grained upwards within the point bar succession, are cross-bedded for some distance above the basal gravel, but cross-laminated in the upper part of the bar succession; the activities of bars, dunes and ripples are recorded. Evenly laminated sands formed on plane beds may occur in any part of the bar though they are perhaps commonest in the upper levels, where they alternate with layers of cross-laminated sand. Beds of silt are thin and confined to the upper part of the bar succession. They are commonly sun-cracked and may also bear rain prints and foot prints. Drifted tree trunks, boughs, twigs and leaves are common in the sands of point bars. The deposits in the upper parts of the bars may include also thin rootlet and peat beds.

Levée and crevasse splay deposits are characterized by evidence of repeated submergence and emergence. Levée deposits generally consist of a vertical alternation, on a scale of centimetres or a few decimetres, of fine or very fine sands with silts and clays, the materials ordinarily carried in suspension by streams. Beds of drifted plant remains are also frequent. The sands commonly have erosional bases, which may also bear sun cracks, and sometimes occupy shallow channels. Usually the sands are cross-laminated, though some prove to be evenly laminated, and others cross-bedded and even pebbly. A vertical grading from coarse up to fine, combined with a sharp base and gradational top, is characteristic of the sand layers found in levées. Where the climate permits plant growth, rootlets penetrate the beds from above. The silt and clay layers vary from massive to delicately laminated, and sometimes include isolated lenses of cross-laminated sand and laminae formed of plant fragments. Crevasse splay deposits are a little coarser than those of levées and commonly reach 3 metres in thickness. They are chiefly sands with a sharp, channelled base and an upward grading from coarse to fine. Very commonly the sand is evenly laminated throughout the greater part of the deposit, cross-lamination being restricted to the uppermost layers. Pebbles and sometimes large blocks of consolidated alluvial silt and clay are commonly found in crevasse splay deposits.

Within the confines of the alluvial ridge there occur cut-off channels. The first sediment to be deposited in an abandoned channel is a plug of channel sand at each end. Once plugged the abandoned channel can receive only the relatively fine sediment carried on to the flood-plain in suspension. The fill of a cut-off channel therefore consists chiefly of silt and clay which has accumulated under virtually lacustrine conditions. Fresh-water molluscs and fish are sometimes preserved in the fill. The process of filling is completed when the cut-off channel is occupied by a swamp or drying mud-flat. The mainly fine grained deposits formed in an abandoned meander loop are curvilinear in plan and about as broad and thick as the original channel was wide and deep.

The deposits laid down in flood-basins are thick sequences predominantly of silt and clay, and rarely contain sand. In temperate and humid climates, where the flood-plain bears a dense plant cover, the deposits are rich in peat layers and horizons of tree roots and stumps preserved in position of growth. Dark colours are characteristic of the sediments. Here and there are lenses of lacustrine clay bearing fish and mollusc remains. In dry climates, flood-basin deposits are brown or red and lack organic debris. Proofs of exposure abound, not only in the form of sun cracks and rain prints, but also in the shape of beds of precipitated salts. Nodular to massive accumulations of calcite, often with some dolomite, are common, as witness the *kunkar* of the Indus alluvial plain and other semi-arid regions. In areas of restricted or internal drainage, sulphates, chlorides and borates are precipitated in flood-basins and preserved between beds of silt or clay.

Although the model of Fig 4.7b covers only a portion of an alluvial ridge, it is easy to see that in an alluvial area of wide extent, such a ridge would form the visible evidence of a rectilinear 'sand' body surrounded by fine grained sediments of flood-basin origin. If avulsion were repeated many times in an alluvial area in which sediment was accumulating, we would have a body of sediment consisting of many such 'sand' bodies embedded in a thick and broad mass of alluvial silt and clay.

4.7 Alluvial Fans and Their Deposits

Alluvial fans are sedimentary bodies shaped like a segment of a cone cut downward from its apex. They are not restricted to arid and semi-arid regions, as is commonly supposed, but are found to occur in every climatic circumstance wherever a water course passes sufficiently rapidly from a steeply dissected terrain to an area of low gradient. Thus alluvial fans abound along mountain and hill fronts and valley sides. Good descriptions of fans and their processes come from the mountains of the south-west USA, Alaska and north-west Canada, the European Alps and the Himalaya.

Compared with other alluvial bodies, alluvial fans are small in size, though when coalescent as an apron along a mountain front they can aggregate to a considerable mass of debris. The smallest fans have a radius of a few hundred metres, while the largest are a few tens of kilometres in radius. An alluvial fan is one to three times larger in area than the drainage basin feeding it, the precise relationship depending on bed rock and climate. The radial profile of alluvial fans is generally concave upward, but is in some cases straight over the whole or a substantial part of the body. The maximum slope of alluvial fans decreases with ascending fan area. For the smallest fans the maximum slope is normally a few degrees, though exceptionally it may be 10°–25°. The larger fans have a maximum slope of less than one degree.

Viewed from the air, alluvial fans show on their surface a radial pattern of channels of various ages of activity, as judged from the lightness of colour of the sediment or the extent and density of the plant cover. Evidently, new sediment is added to alluvial fans along narrow radial bands, sometimes terminating in lobe-shaped zones, the area of deposition shifting about on the fan as time passes. Three processes—stream flow, debris flow, and mud flow—are responsible for fan deposition, and the balance between them is determined chiefly by climate and to only a small extent by bed rock.

Stream flows predominate in areas of moderate year-round rainfall. Transport and deposition occur in powerful, braided perennial streams which shift about on the fan as one area after another is built up to a disadvantageous extent. Stream flow deposits are chiefly well bedded and well sorted gravels, sandy gravels and sands. Particle orientation is good. Very commonly the sandier beds are cross-bedded, while the gravelly ones fill channels.

Debris flows and mud flows are commonest in relatively dry regions of scattered but intense rainfall; mud flows are also favoured by cold conditions. The relatively high sediment concentrations encountered in debris and mud flows can, at least in hot dry regions, be attributed to the absorption of water from the flow by the parched body of the fan. A debris flow differs from a stream flow in that the flow is of limited duration and the sediment concentration is higher; both types of flow are relatively swift. Debris flows deposit material

on steep-sided levées along the lateral margins of the flow and in lobe-shaped mounds at the termination of the flow. Debris flow deposits are chiefly gravelley sands and sandy gravels that are less well sorted and bedded than stream flow deposits. The particles are commonly well oriented but are equally often chaotic. Vertical grading from coarse up to fine is common in debris flow deposits. Mud flows are, on the other hand, highly concentrated mixtures of clay, silt, sand and gravel, often with the fine grades predominant. Their production is favoured by the presence of clay-bearing rocks in the source area. Depending on bed slope and viscosity, mud flows travel rapidly as surges down fan channels, or glide along at a walking pace or less. Their deposits are extremely poorly sorted, ordinarily chaotic, and generally without vertical grading. In hot regions they become deeply cracked on drying.

Alluvial fan sediments rapidly become finer grained with increasing distance from the fan apex. The deposits, invariably of local rocks, pass downslope into normal flood-plain sediments or into playa or lake beds depending on circumstances.

4.8 Ancient Alluvial Sediments

Sedimentary rocks of alluvial origin present few problems as regards recognition, and are well represented in the geological record. Many cases have been studied in great detail and a few, such as the Catskill (Devonian) of the Appalachian region, are now classics of the sedimentological literature.

Ancient alluvial fan and braided stream deposits are represented by thick conglomerates of Devonian age in western Norway and eastern Greenland, and again by conglomerates and sandstones of Triassic age preserved in large basins on the eastern seaboard of the USA. The latter basins are partly fault-bounded, and it appears that the conglomerates and sandstones accumulated against active discolations. Towards the basin centres the conglomerates and sandstones pass into lacustrine silts and clays.

In Great Britain and north-west Europe and in the Appalachian region of North America there occurs the Old Red Sandstone of Devonian age, consisting of sandstones and siltstones, mainly red in colour, with generally minor conglomerates. This facies originated on coastal plains of alluviation of huge extent that bordered the

Caledonian mountain chain, during Devonian times in its final paroxysms of folding. The alluvial plain was in turn flanked by the sea. The plain was the product of many rivers. Near the mountains, the rivers and streams had steep slopes and were probably braided, for the deposits are chiefly sandstones and conglomerates with but few siltstones indicative of extensive flood-plain deposition. Channel fills abound, however. Nearer the sea the rivers of the Old Red Sandstone alluvial plain seem to have meandered within alluvial ridges. The deposits are marked by a distinctive vertical repetition of different rock types on a scale measured in metres or a few tens of metres. The basic unit of this repetition comprises in order of upward appearance: (i) conglomerate resting on an eroded and sometimes channelled surface, (ii) cross-bedded, evenly laminated and cross-laminated sandstone, (iii) alternations of sandstone and siltstone with proofs of exposure and sometimes beds of concretionary limestone resembling *kunkar* and, (iv) thick siltstones often with concretionary limestones. This unit, repeated many times over in a vertical sequence through the deposits, is remarkably like the succession of sediments found in the flood-plain of a modern meandering stream. The repetition of the pattern in the rocks could well record the repeated avulsion of the Devonian rivers.

Of great economic importance in Europe and the western USSR, and again in the interior regions of the USA, are coal-bearing rocks of Carboniferous age. A considerable part of these beds is of alluvial origin, though marine rocks are also present. In the alluvial sections, deep channels filled with sandstone are of frequent occurrence. The coals seem to represent flood-plain and coastal swamps.

READINGS FOR CHAPTER 4

Discussions of fluvial processes and alluvial geomorphology can be found, variously emphasized, in the following general works:

ALLEN, J. R. L. 1965. 'A review of the origin and characteristics of recent alluvial sediments.' *Sedimentology*, **5**, 89–191.

MORISAWA, M. 1968. *Streams: their Dynamics and Morphology.* McGraw-Hill Book Co., New York, 175 p.

LEOPOLD, L. B., WOLMAN, M. G. and MILLER, J. P. 1964. *Fluvial Processes in Geomorphology.* W. H. Freeman and Co., San Francisco, 522 p.

RAUDKIVI, A. J. 1967. *Loose Boundary Hydraulics.* Pergamon Press, Oxford, 331 p.

Sundborg, Å. 1956. 'The River Klarälven: a study of fluvial processes.' *Geogr. Annlr*, **38,** 127–316.

The books of Morisawa and Leopold illustrate the fairly common tendency amongst geomorphologists to emphasize the role of rivers in erosional landscapes at the expense of depositional ones. Allen's review, however, deals fairly completely with the depositional landscapes created by streams. The book of Raudkivi, written from the standpoint of the engineer, is a valuable survey of channel form and process and of sediment transport in streams. Sundborg's article covers similar ground, but the attitude is definitely that of a morphologist.

A host of articles exist on the deposits of modern streams. Unfortunately, the emphasis in research has fallen on channel sediments, which are more attractive than those of the flood-plain, and on sand-bed rather than gravel-bed streams. The following list is far from comprehensive, though it includes the classic and still pertinent works of Fisk and Jahns, and rather inevitably concentrates on sand-bed rivers:

Davies, D. K. 1966. 'Sedimentary structures and subfacies of a Mississippi River point bar. *J. Geol.*, **74,** 234–239.

Doeglas, D. J. 1962. 'The structure of sedimentary deposits of braided rivers.' *Sedimentology*, **1,** 167–190.

Fahnestock, R. K. 1963. 'The morphology and hydrology of a glacial stream, White River, Mount Rainier, Washington.' *Prof. Pap. U.S. geol. Surv.*, **422-A,** 70 p.

Fisk, H. N. 1944. *Geological investigation of the alluvial valley of the Lower Mississippi River*. Mississippi River Commission, Vicksburg, Miss., 78 p.

Frazier, J. and Osanik, A. 1961. 'Point-bar deposits, Old River Locksite, Louisiana.' *Trans. Gulf-Cst Ass. geol. Socs*, **11,** 121–137.

Harms, J. C. and Fahnestock, R. K. 1965. 'Stratification, bed forms and flow phenomena (with an illustration from the Rio Grande).' *SEPM Spec. Publ.*, **12,** p. 84–115.

Jahns, R. H. 1947. 'Geological features of the Connecticut Valley, Massachusetts, as related to recent floods.' *Wat.-Supply Irrig. Pap., Wash.*, **996,** 158 p.

McKee, E. D., Crosby, E. J. and Berryhill, H. L. 1967. 'Flood deposits, Bijou Creek, June 1965.' *J. sedim. Petrol.*, **37,** 829–851.

Simons, D. B., Richardson, E. V. and Nordin, C. F. 1965. Sedimentary structures generated by flow in alluvial channels. *SEPM Spec. Publ.*, **12,** p. 34–52.

Williams, P. F. and Rust, B. R. 1969. 'The sedimentology of a braided river.' *J. sedim. Petrol.*, **39,** 649–679.

Although alluvial fans are of world-wide distributions, our knowledge of them is greatest from hot dry regions. The following are representative:

Beatty, C. B. 1963. 'Origin of alluvial fans, White Mountains, California and Nevada.' *Ann. Ass. Am. Geogr.*, **53,** 516–535.

Blissenbach, E. 1964. 'Geology of alluvial fans in semi-arid regions.' *Bull. geol. Soc. Am.*, **65,** 175–190.

BLUCK, B. J. 1964. 'Sedimentation of an alluvial fan in southern Nevada.' *J. sedim. Petrol.*, **34,** 395–400.

BULL, W. B. 1964. 'Alluvial fans and near-surface subsidence in western Fresno County, California.' *Prof. Pap. U.S. geol. Surv.*, **437-A,** 70 p.

HOOKE, R. LeB. 1967. 'Processes on arid-region alluvial fans.' *J. Geol.*, **75,** 438–460.

WINDER, C. G. 1965. 'Alluvial cone construction by alpine mudflow in a humid temperate region.' *Can. Jl Earth Sci.*, **2,** 270–277.

Many rock successions have been interpreted as alluvial in origin, but those of Devonian or Carboniferous ages in Europe and North America are perhaps the best known. The following papers are representatives of recent work:

ALLEN, J. R. L. 1964. 'Studies in fluviatile sedimentation: six cyclothems from the Lower Old Red Sandstone, Anglo-Welsh Basin.' *Sedimentology*, **3,** 163–198.

ALLEN, J. R. L. 1965. 'The sedimentation and palaeogeography of the Old Red Sandstone of Anglesey, North Wales.' *Proc. Yorks. geol. polytech. Soc.*, **35,** 139–185.

ALLEN, J. R. L. 1970. 'Studies in fluviatile sedimentation: a comparison of fining-upwards cyclothems, with special reference to coarse-member composition and interpretation.' (To be published in *J. sedim. Petrol.*)

BLUCK, B. J. and KELLING, G. 1962. 'Channels from the Upper Carboniferous Coal Measures of South Wales.' *Sedimentology*, **2,** 29–53.

MOODY-STEWART, M. 1966. 'High- and low-sinuosity stream deposits, with examples from the Devonian of Spitsbergen.' *J. sedim. Petrol.*, **36,** 1102–1117.

POTTER, P. E. 1963. 'Late Paleozoic sandstones of the Illinois Basin.' *Rep. Invest. Ill. St. geol. Surv.*, **217,** 92 p.

CHAPTER 5

Waves, Tides, and Oceanic Circulations: Shallow-Marine Deposits

5.1 General

Most of the particles of debris formed by the weathering of the land eventually succeed in reaching the seas and oceans, which at present comprise approximately 71 per cent of the area of the earth, that is, 350×10^6 km^2. Whilst the land masses act as a source of sediment, the oceans and seas are a colossal sink, with rivers serving as the chief conveyance system linking the two. Material travels from sink back to source, completing the cycle, only when the sediments which have accumulated in a marine basin are folded and thrust up above the sea to form new land with hills, mountains and rivers.

The particles that reach the seas and oceans find themselves in a remarkably complicated environment, of which only the physical aspects are here of concern. In this environment currents of four main kinds exist as destructive, dispersive and constructive agents: wind-wave currents, tidal currents, oceanic circulatory currents, and turbidity currents. In these agents are stored enormous amounts of energy, in some cases because of the large quantity of matter involved in the fluid flow, and in others because of the relatively high velocity of the current.

Currents of the first three kinds are most effective in the shallower parts of the sea, particularly on the continental shelves whose depth is generally less than 200 metres. At the shore itself, and in estuaries and tidal flats, and lagoons and bays backward of coastal barriers, only wave and tidal currents are significant. In waters deeper than about 200 metres, oceanic circulations and turbidity currents are chiefly responsible for the dispersal and deposition of clastic sediment, for at these depths wave and tidal movements are negligible. Turbidity currents are, however, so important for deep-sea sedimentation that we shall reserve discussion of them until Ch. 6.

Our purpose in the present chapter will be briefly to examine sedimentation under the influence of waves, tides and oceanic circulations. These agencies are of enormous concern to the geologist dealing with the rock succession, as marine deposits, some of which are bioclastic, are commoner than sediments of other origins, and amongst marine deposits those of shallow-water genesis rank high in abundance. Shallow-water marine sediments are, moreover, important in many parts of the world as oil or gas reservoirs, so there are strong economic motives for a consideration of processes acting at sea. Other economic stimuli, no less important, are afforded by the needs of fisheries and the coast protection services.

5.2 Wave Theory

The waves we are interested in occur as more or less regularly spaced hummocks and hollows of many different scales on the air-sea interface. These waves are surface waves, as opposed to internal waves which sometimes occur on the interface between two water bodies of slightly different density, one above the other. The waves of interest are also mostly progressive, for the profile is found to move relative to the water, though some natural waves must be classified as standing (the tide on occasions), because the profile is seen merely to oscillate in one place relative to the medium.

For ideal progressive waves on infinitely deep water, we can write with g as the acceleration due to gravity

$$\lambda = \frac{g}{2\pi} T^2, \tag{5.1}$$

and

$$c = \left(\frac{g\lambda}{2\pi}\right)^{\frac{1}{2}}, \tag{5.2}$$

in which, as shown in Fig 5.1, the wavelength λ is the distance between corresponding points on the profile, and the wave period T

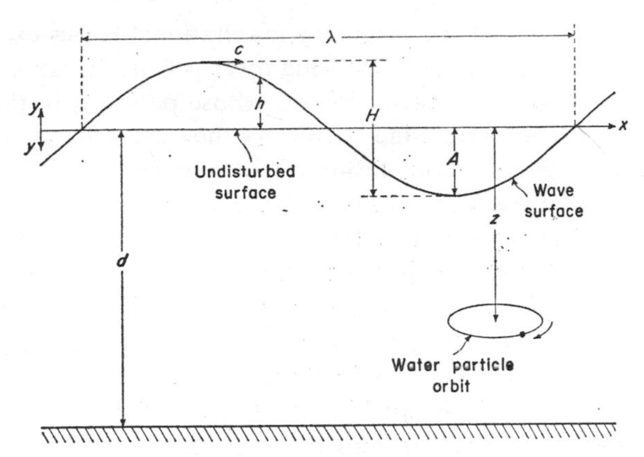

Fig 5.1 Definition diagram for progressive surface waves.

is the time required for corresponding points to pass a stationary observer. The phase speed, or simply the celerity, denoted by c, is the speed at which an initial point on the wave profile passes the observer. The most convenient initial point by which to measure the wave speed is either the wave crest or the trough; on the surface of the sea the waves are propagated in a direction normal to their crests. The vertical distance between crest and trough is the wave height H which in the case of a sinusoidal wave is twice the wave amplitude A. It is also useful to write

$$\eta = 2\pi/T, \tag{5.3}$$

$$\kappa = 2\pi/\lambda, \tag{5.4}$$

and

$$\sigma = H/\lambda, \tag{5.5}$$

where η and κ are the radian frequency and radian wave number respectively, and σ is the wave steepness. Then from Eqs. (5.1–4)

$$c = \lambda/T = \eta/\kappa. \tag{5.6}$$

By invoking the period or frequency, we can classify progressive surface waves according to the forces chiefly responsible for generating them and the forces chiefly active in the attempt to flatten the water surface. The longest period waves are those represented by the tides, about which more will be said below. Their period is either about 24 hours or about 12 hours, depending on whether the tide is diurnal or semi-diurnal. Tidal 'waves' are generated by the action on the seas and oceans of the planetary gravitational forces exerted by the sun and moon. The chief restoring force is Coriolis force, arising from the rotation of the earth. Waves whose period is in the range $300 > T > 0{\cdot}01$ secs are wind-waves, as they are generated on the sea surface by wind action, though in a manner far from simple. The smaller of these waves ($T < 0{\cdot}1$ secs) are known as capillary waves, for surface tension is the chief restoring force. Capillary waves are not geologically interesting, though physically important. The earth's gravity is the chief restoring force in the case of the larger waves ($T > 0{\cdot}1$ secs), which are therefore called gravity waves. Real gravity waves are found in three states. They are called sea when the wind is actively working on the water surface, and swell when they have travelled beyond the influence of the wind. Surf is the name given to gravity waves breaking up on the shore. Breaking waves as seen on sloping beaches are of four main kinds—spilling, plunging, collapsing and surging—which differ in cross-sectional appearance.

An important classification of progressive surface waves depends on the ratio of water depth d to wavelength λ_0, the subscript denoting that the wavelength is measured in or calculated for deep water (Fig 5.1). If d/λ_0 is small, the wave is called a shallow-water or long wave, whereas if d/λ_0 is large, the disturbance is called a deep-water or short wave. These conditions, together with a range of intermediate ones, can be specified as follows, where the ratio d/λ_0 is the relative depth:

$d/\lambda_0 \geqslant 0{\cdot}5$ deep water;

$0{\cdot}05 < d/\lambda_0 < 0{\cdot}5$ intermediate water;

$d/\lambda_0 \leqslant 0{\cdot}05$ shallow water.

Thus we imply that a wave *remains of constant period* as it travels from deep to shallow water, though its speed, wavelength and height will certainly vary as the water depth changes.

Since real waves cover a very broad spectrum in terms of period, steepness, and relative depth, it is hardly surprising that several different mathematical theories of progressive surface waves have been advanced. The validity of these different theories is determined by whether or not H/d and H/λ_0 are sufficiently small compared with unity and by the relative magnitude of d/λ_0. All the theories assume frictionless conditions. In the small-amplitude or Airy theory, which is the simplest thus far developed, wave amplitude is considered negligibly small compared with water depth. This theory, by using suitable approximations, covers all values of d/λ_0, since we can vary λ_0 whilst still keeping H negligibly small compared with d. The other theories, of which there are three main ones, are mathematically more complicated and are developed on the supposition that the wave amplitude is not negligibly small compared to water depth. Of these the Stokesian theory is valid for $d/\lambda_0 \geqslant 0{\cdot}1$ and the solitary wave theory for $d/\lambda_0 \leqslant 0{\cdot}02$. The finite-amplitude theory valid for $0{\cdot}1 > d/\lambda_0 > 0{\cdot}02$ is for cnoidal waves. Within the general framework of finite-amplitude theory there exist also the Gerstner-Rankine theory for deep-water waves and Crapper theory for capillary waves. Only the small-amplitude and the Stokesian theories are further considered below.

According to small-amplitude theory, the wave profile is given by

$$h = A \sin(\kappa x - \eta t), \tag{5.7}$$

where, as shown in Fig 5.1, h is the difference of elevation between the profile and the undisturbed water surface, x is distance along the horizontal coordinate in the direction of wave propagation, and t is time. If x is held constant we obtain from Eq. (5.7) the change of level of the water surface at a point with time, and if t is held constant, the profile of the surface at an instant some time after the disturbance started. The general equations for celerity and wavelength are

$$c^2 = \frac{g}{\kappa}\tanh(\kappa d), \tag{5.8}$$

and

$$\lambda = \frac{gT^2}{2\pi}\tanh(\kappa d). \tag{5.9}$$

Tables and graphs are widely available showing how celerity and wavelength vary with water depth. By inserting into Eqs. (5.8) and (5.9) the asymptotes of the hyperbolic functions, we obtain the following approximation for short (deep-water) waves:

$$\left.\begin{aligned} c_0 &= \frac{g}{2\pi}T \\ \lambda_0 &= \frac{g}{2\pi}T^2 \end{aligned}\right\}, \tag{5.10}$$

since $\tanh \kappa d \rightarrow 1$ as $d/\lambda_0 \rightarrow \infty$, the subscript indicating the deep-water value. Similarly, for long (shallow-water) waves

$$c = \sqrt{(gd)}, \tag{5.11}$$

since $\tanh \kappa d \rightarrow \kappa d$ as $d/\lambda_0 \rightarrow 0$. The full equations must, however, be used for intermediate water conditions ($0{\cdot}05 < d/\lambda_0 < 0{\cdot}5$). It will be seen that for deep-water conditions, the celerity is independent of depth, whilst for shallow-water waves the celerity is dependent only on depth.

The small-amplitude theory also says that the water particles beneath a progressive wave describe closed orbits represented by the simplified equation

$$\left(\frac{x'}{X}\right)^2 + \left(\frac{y'}{Y}\right)^2 = 1, \tag{5.12}$$

in which x' and y' are the horizontal and vertical displacements, respectively, of the particle from its mean position. The equation is that of an ellipse of major (horizontal) semi-axis X and minor (vertical) semi-axis Y. Putting z (negative) as distance measured below the undisturbed water surface to the geometrical centre of the particle orbital path (Fig 5.1), we have

$$x' = A \frac{\cosh \kappa(d+z)}{\sinh \kappa d} \cos(\kappa x - \eta t), \tag{5.13}$$

and

$$y' = A \frac{\sinh \kappa(d+z)}{\sinh \kappa d} \sin(\kappa x - \eta t). \tag{5.14}$$

In the general case

$$\left.\begin{aligned} X &= A \frac{\cosh \kappa(d+z)}{\sinh \kappa d} \\ Y &= A \frac{\sinh \kappa(d+z)}{\sinh \kappa d} \end{aligned}\right\}, \tag{5.15}$$

but we can make a number of approximations, writing for deep-water conditions

$$\left.\begin{aligned} X &= Ae^{\kappa z} \\ Y &= Ae^{\kappa z} \end{aligned}\right\}, \tag{5.16}$$

and for shallow-water conditions

$$\left.\begin{aligned} X &= A\frac{1}{\kappa d} \\ Y &= A\frac{\kappa(d+z)}{\kappa d} \end{aligned}\right\}, \tag{5.17}$$

in which e is the base of natural logarithms.

The velocity of water particles as they traverse their orbits is also given by the theory, and we can write for the horizontal component u of the particle velocity

$$u = \frac{Ag\kappa}{\eta}\frac{\cosh\kappa(d+z)}{\cosh\kappa d}\sin(\kappa x-\eta t). \tag{5.18a}$$

The maximum value attained by u, as $z \to d$, is of considerable interest in connection with the behaviour of sediment particles at the bottom of water affected by surface waves. For this purpose, Eq. (5.18a) as the general relation between u, x, z and t may be simplified to read

$$u_{\max} = \frac{A\eta}{\sinh \kappa d}, \tag{5.18b}$$

in which $u_{\max}$ is the maximum horizontal velocity attained by water particles immediately above the bottom at a depth d. The equation is valid for deep-water conditions, when $u_{\max}$ proves to be zero at the bottom, and also for intermediate conditions, when the maximum bottom velocity is found to take non-zero values. For a wave travelling under shallow-water conditions, Eq. (5.18a) reduces still further to

$$u_{\max} = \frac{A}{d}\,c, \tag{5.18c}$$

where the celerity is the shallow-water value given by Eq. (5.11). For example, a wave of period equal to 10 seconds and amplitude equal to 20 cm would give rise at the bottom of water 5 metres deep to a current whose maximum horizontal velocity was nearly 30 cm/s, amply sufficient to entrain quartz sand.

Inspection of these equations will reveal their physical meaning, which is summarized in Fig 5.2. For deep-water conditions the particle orbits are circles whose radius decreases exponentially downwards until, at a depth of $\lambda/2$, the motion becomes negligibly small.

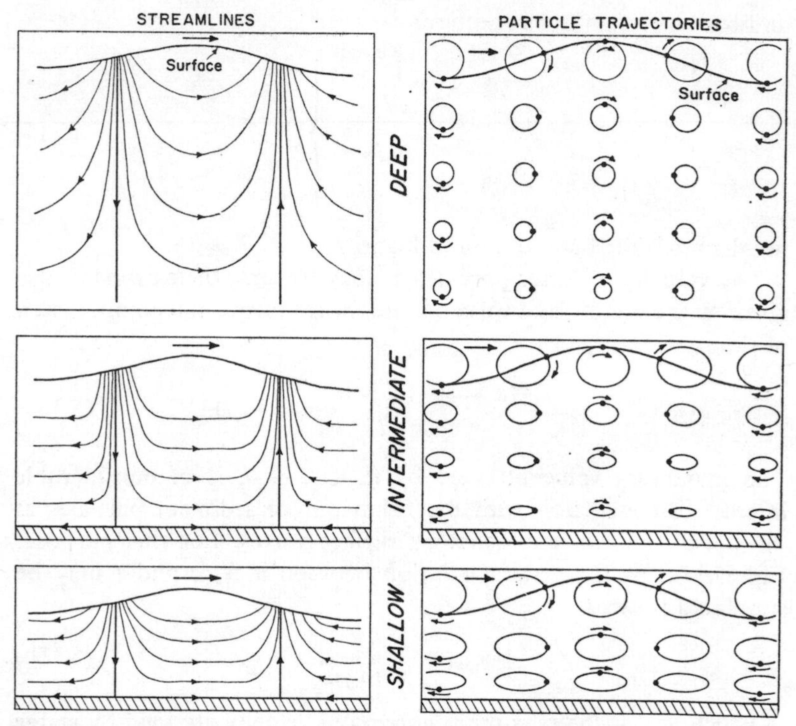

Fig 5.2 Streamlines and water-particle trajectories for progressive waves advancing from left to right over the surface of deep, intermediate or shallow water relative to the wavelength.

The depth equal to $\lambda/2$ is called wave-base for the deep-water condition. It is the greatest depth at which the surface waves can influence the bottom. Under shallow-water conditions, however, the orbits are ellipses which remain of constant horizontal semi-axis with increasing depth but become progressively flatter downwards, the maximum horizontal component of the orbital velocity remaining unchanged. For conditions of intermediate depth, the orbits are ellipses of constant eccentricity whose size decreases progressively downward but is finite at the bottom itself. The streamline patterns of Fig 5.2 are those for an instant of time, and it should be noted that the patterns for shallow and intermediate water include singularities on the bottom. Of course, these patterns, together with their singularities, oscillate over a path parallel to the direction of wave travel.

One final result of small-amplitude theory is an expression for the energy of a simple harmonic wave. Such a wave has potential energy by virtue of the deformation of the free surface and kinetic energy arising from motion of the water. In our small-amplitude case the average potential energy density is equal to the average kinetic energy density. The total energy per unit surface area is the sum of these two quantities, given by

$$E = \tfrac{1}{2}\rho g A^2, \tag{5.19}$$

where E is the average energy per unit surface area, and ρ is the fluid density. The energy is proportional to the square of the wave amplitude, but it is transported at a speed less than the wave celerity. Looking at the natural environment as a whole, the largest amounts of energy are stored in the tidal 'waves' and in gravity waves of $T \approx 10$ secs.

The finite-amplitude theories afford a better approximation to the truth than the small-amplitude theory. The Stokesian theory states that the water particle orbits are not in fact closed. This means that a mass movement of water, in the direction of wave propagation, occurs beneath a train of progressive waves. The variation of the mass transport velocity U_m with depth below the surface is given by

$$U_m = \left(\frac{\pi H}{\lambda}\right)\frac{c}{2}\frac{\cosh 2\kappa(d+z)}{\sinh^2 \kappa d}. \tag{5.20}$$

Thus if the waves are moving towards some obstacle—a line of cliffs or a beach—continuity demands that the forward mass transport arising from the wave motion is compensated for by an equal mass return flow. Another major result of Stokesian theory concerns the conditions for wave breaking. When breaking occurs, the two faces of the wave make an angle of 120° at the crest, and the water particle velocity at the crest is just equal to the wave celerity. For if the particle velocity exceeds the celerity, the water 'overtakes' the wave and the wave begins to spill. The condition for breaking is

$$\frac{H}{T^2} \geqslant 0{\cdot}875. \tag{5.21}$$

Wave reflection, refraction, and diffraction are important in determining the movement of sediment in shallow water close to shore and therefore in shaping coastlines of accretion. These matters can be treated in much the same way as geometrical optics, though the theory of diffraction presents difficulties. Wave reflection from breakwaters and steep beaches is a commonplace. The reflected wave is seen to make the same angle with the obstacle as the oncoming one (Fig 5.3), though because of dissipation it is unlikely to be as high. Since, by Eq. (5.8), the wave celerity depends on water depth for all but deep-water conditions, the refraction of waves passing into shoaling water can be treated by Snell's law of geometrical optics. Thus, waves obliquely approaching a straight coastline that shelves away into deeper waters are found to bend in towards the depth contours (Fig 5.3). Quite complicated patterns arise when waves are refracted around shelving islands, and above troughs or ridges (Fig 5.3). Waves become diffracted, or spun round, on passing the edge of an obstacle, such as a breakwater or sand spit, and some of their energy is transmitted into areas of wave shadow (Fig 5.3).

Most waves eventually reach the shore, where they break up to

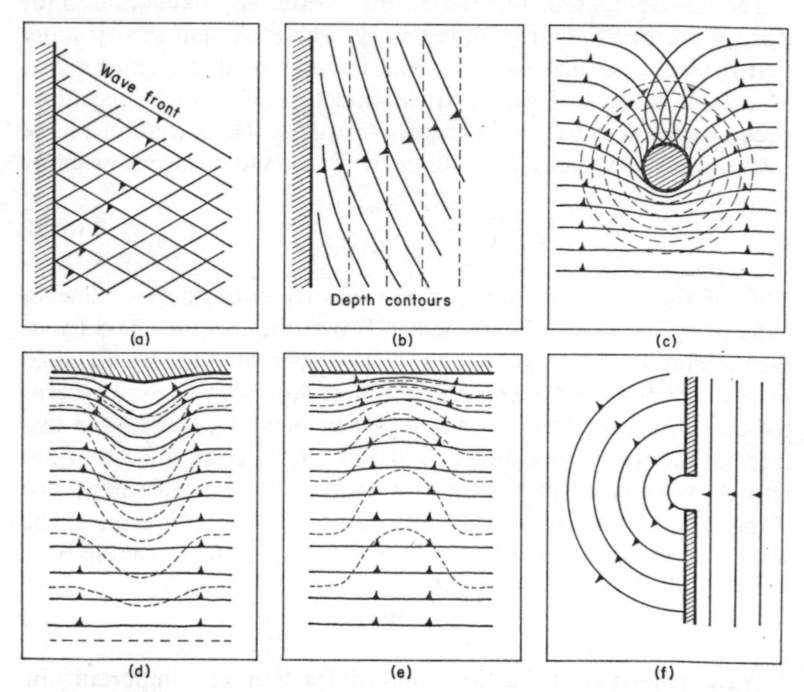

Fig 5.3 Idealized behaviour of progressive waves at obstructions. (a) Reflection from a steep cliff. (b) Refraction in shoaling water. (c) Refraction around an island sited on a radially deepening bottom. (d) Refraction above a submerged ridge. (e) Refraction above a submerged hollow. (f) Diffraction at an inlet.

form surf. When the waves approach the shore obliquely, some of the fluid in motion in the waves is transferred into longshore currents that flow parallel to the coast landward of the breaker zone. Theories of longshore currents based on momentum or energy considerations are unsuccessful, as they depend on an untenable assumption. But considerations of conservation of mass lead to a valid theory in fair agreement with observation. Setting U_L as the mean velocity of the longshore current,

$$U_L = KgST \sin 2\theta, \tag{5.22}$$

where K is a dimensionless coefficient taken to be unity, S is the beach slope, and θ is the angle between the wave crest and the breaker line at the point of wave-breaking. It is at first sight surprising that Eq. (5.22) does not contain the wave height. However, the mass rate of flow must strongly depend on wave height, since the height controls the cross-sectional area of the longshore current by determining the breaker point.

Before leaving wave theory, a little should be said concerning, firstly, the differences between real and ideal waves and, secondly, wave generation and decay.

The most casual observer knows that even under the best of natural conditions, the waves on the surface of the sea at any one time in a locality are of a whole range of sizes. A spectrum of waves is in fact present. By using special instruments, the waves passing a station can be described in terms of frequency distributions of observed period, height and wavelength. A two- or three-fold range of variation of each of these properties is to be expected even for sheltered conditions. When conditions are stormy, with the wind very gusty, a much broader range of variation is ordinarily observed. Averages can be calculated from the frequency distributions, but the average period, height, and wavelength as obtained from spectra of real waves do not have quite the same meaning as the corresponding, but single-valued, terms used in a theoretical treatment. In this connection, and since energy is proportional to amplitude, it is of interest to know how height is distributed amongst real ocean waves. An analysis of more than 40000 observations shows that 45 per cent of ocean waves are less than 1·2 metres in height, 80 per cent are less than 3·6 metres high, while only 10 per cent exceed a height of 6 metres. The largest wind-wave ever reliably recorded at sea was approximately 34 metres high.

Wave generation by wind is one of the most complicated, and also misunderstood, of all questions in fluid dynamics, but the chief facts concerning the problem are as follows. A short time after the wind starts to blow over still water, the water surface changes from glassy-smooth to ruffled, the ruffles generally being distributed in patches that move with, but slower than, the wind. The ruffled areas in fact consist of waves of small height and wavelength. Some of the waves are capillary waves a centimetre or two in wavelength, though others are several centimetres in wavelength and form rhombic patterns. With the continued blowing of the wind, the waves grow noticeably in height and wavelength. For a given wind and distance of travel of the waves under the influence of the wind—the fetch distance—there arises a characteristic wave spectrum. A theoretical analysis of the problem shows at once that waves cannot arise in response to a tangential wind force. Therefore the wave-generating force must be a *normal dynamic pressure force*, presumably varying in magnitude in space and time. A turbulent wind is at once implied, since a steady laminar flow over a perfectly smooth flat surface cannot exert a varying pressure force.

Several different mechanisms of energy transfer from wind to wave are involved in wave-generation, each being most effective over a particular narrow range of wave celerity. The initial growth of waves as a wind begins to blow over the water surface is due to a resonant or forced response of the free surface to the ever-changing turbulent wind eddies convected past the surface. The waves thus generated are small and commonly at steep angles to the mean wind. They grow linearly with time, and the growth rate is largest for those whose celerity equals the eddy convection velocity. But as the waves grow larger, a coupling arises between the wave system and the wind flow over it. Essentially, the wave form exerts a slight but significant influence on the wind, which in turn affects the wave, and so on, the two effects keeping in step. The coupling is expressed physically as a normal pressure component in phase with the wave slope. With coupling between wave system and wind, energy transfer can occur in at least three ways, none of which can be simply described. The extent of transfer in one mechanism depends on the gradient and curvature of the mean wind velocity profile. Waves in the range $10U_* < c < 20U_*$, where U_* is the wind shear velocity, grow most rapidly under this mechanism. A second transfer mechanism involves intermittent fluctuations of the turbulence of the wind, and affects most strongly those waves for which $c \approx 18U_*$. The third mechanism of energy transfer is connected with a viscous instability of the laminar sublayer of the wind. Waves for which $c < 10U_*$ are most strongly affected by the mechanism.

5.3 Tidal Theory

The tide is a rhythmical rise and fall of the sea surface accompanied by horizontal currents, very strong in some places but weak in others. This oscillation of the sea arises because the moon and sun exert an attraction on the mobile ocean water that covers the greater part of the surface of the earth. The tide-generating forces are complicated, however, owing to the rotation of the earth and sun about their axes, the elliptical orbits of the earth and moon, the different attitudes of the orbital planes, and the changes of declination of the sun and moon throughout their orbital cycles, amongst other factors. When the total tide-generating force is analysed into its component parts, seven main components, each giving rise to a partial tide, can be recognized. The most important is the principal lunar semi-diurnal component given the symbol M_2. Putting the amplitude of M_2 as 100, we can list these seven as in Table 5.1. It will be noticed that three components are diurnal and that the M_2-component is not outstandingly important.

A first approximation to tidal phenomena is given by the equilibrium theory of the tide. We assume a spherical earth completely covered by a frictionless ocean, and examine the equilibrium

TABLE 5.1

Principal tidal components

Component	Symbol	Approximate period	Ratio M_2: 100
Principal lunar	M_2	Semi-diurnal	100·0
Principal lunar (diurnal)	O_1	Diurnal	41·5
Larger lunar elliptic	N_2	Semi-diurnal	19·2
Principal solar (diurnal)	P_1	Diurnal	19·4
Principal solar	S_2	Semi-diurnal	46·6
Luni-solar	K_1	Diurnal	58·4
Luni-solar (semi-diurnal)	K_2	Semi-diurnal	12·7

shape assumed by this ocean under all the forces considered to be acting. The tide-producing body may be either the moon or sun, and the tide is observed as the fluctuation of water level along a complete line of constant latitude. Then the tide-producing force acting on unit mass of water is

$$F = -\frac{3}{2}g\frac{M}{M_E}\left(\frac{r}{L}\right)^3 \sin 2\theta, \tag{5.23}$$

where M is the mass of the tide producing body, M_E is the mass of the earth, r is the radius of the earth, L is the distance between the centre of the earth and the centre of the tide-producing body, and θ is the angle subtended at the centre of the earth by a point on the earth's surface relative to the line joining the centres of the earth and the tide-producing body. By requiring the ocean surface to be normal to the local resultant force, the deformation of the water surface becomes

$$h = \frac{r}{4}\frac{M}{M_E}\left(\frac{r}{L}\right)^3 (3 \cos \theta + 1), \tag{5.24}$$

where h is the deviation from the undisturbed level. This equation represents a prolate spheroid with major axis directed toward the tide-producing body (Fig 5.4). With the moon as the tide-producing

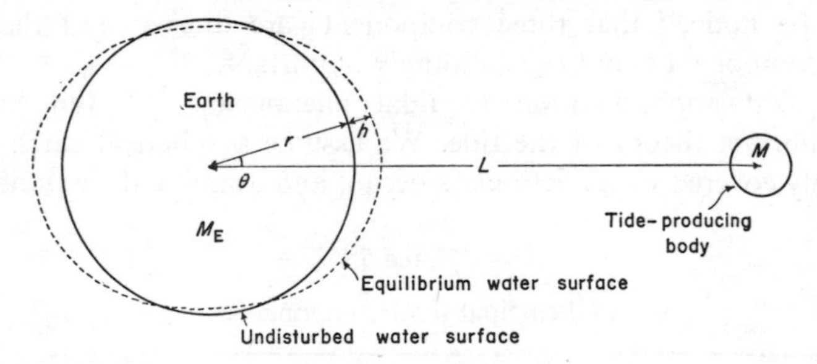

Fig 5.4 Definition diagram for the equilibrium theory of the tides.

body, the maximum deviation from the undisturbed surface is $h = +35{\cdot}75$ cm for $\theta = 0°$ or $180°$, while the minimum deviation is $h = -17{\cdot}86$ cm for $\theta = 90°$. Calculations for the sun yield a tidal range which is a little less than half that obtained for the moon; the relatively larger mass of the sun is more than offset by its relatively greater distance from the earth. Many observed tidal phenomena can be explained by the equilibrium theory if we now assume that the equilibrium ocean surface remains stationary as an observer traverses on a rotating earth a line of constant latitude.

The theory predicts, for example, that because of the declination of the axis of the moon, the tide in low latitudes is semi-diurnal but in high latitudes is diurnal. Further, it predicts the slightly larger tidal range experienced in low latitudes. If we consider in its light the combined tide-generating effect of moon and sun, we obtain the observed bi-monthly phenomena known as neap and spring tides, and also the roughly two-fold increase of tidal range from neaps to springs. But the equilibrium theory does not involve the confining effects of ocean and sea basins, nor frictional losses at the sea bed and the Coriolis force. Coriolis force, the gyroscopic effect due to rotation of the earth, has a particularly strong influence on tidal motion, as we shall now see.

Consider a tidal wave advancing down an infinitely long but frictionless channel on a rotating earth. The effect of rotation is to generate transverse currents in addition to variations of tidal displacement and longitudinal current. The coordinate system for this problem has the x-direction parallel to the channel and the z-direction transversely across the channel. From the wave theory above, but using the cosine rather than the sine function, the tidal displacement becomes

$$h = A \cos(\eta t - \kappa x), \tag{5.25}$$

where A is the tidal amplitude dependent on z. The tidal velocities u and v, along and across the channel respectively, are assumed to be

$$\left.\begin{aligned} u &= U \cos(\eta t - \kappa x) \\ v &= V \sin(\eta t - \kappa x) \end{aligned}\right\}, \tag{5.26}$$

where U and V are maximum velocities again dependent on z. Then by Eq. (5.11)

$$c = \eta/\kappa = \sqrt{(g\bar{d})}, \tag{5.27}$$

in which c is the celerity of the tidal wave and $\bar{d}$ is the mean depth of the channel. Eq. (5.26) can be rewritten as the equation of an ellipse of major semi-axis U and minor semi-axis V, thus

$$\left(\frac{u}{U}\right)^2 + \left(\frac{v}{V}\right)^2 = 1. \tag{5.28}$$

Hence the tidal current varies in magnitude and direction with time. If, using Eq. (5.28), we plot out the tidal current vector at a fixed station at equal time-intervals over one tidal period, we obtain the elliptical pattern of Fig 5.5. Real tidal currents are rather more complicated than this simple figure, but almost all display in their pattern one or more ellipses. The currents rotate clockwise in the Northern Hemisphere but anticlockwise in the Southern Hemisphere.

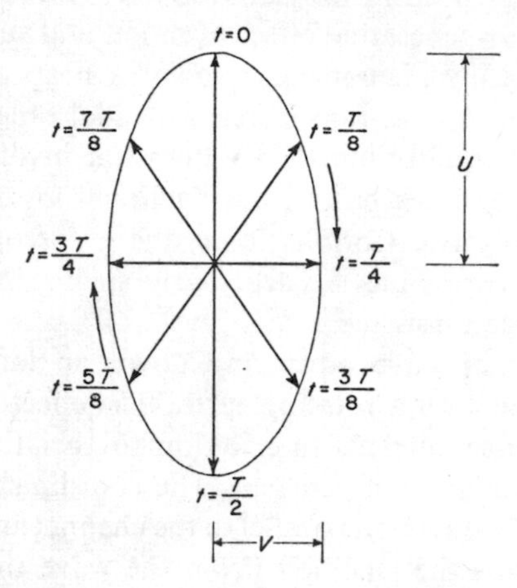

Fig 5.5 Tidal current vectors for different fractional portions of the tidal period T, satisfying the condition $U = 2V$.

Turning to the tidal motion in basins whose width is similar to their length, we again note the important influence of Coriolis force. A simple experiment, providing an analogy, will illustrate the results of this influence. Partly fill a large circular bowl with water and leave aside until the water is still. Upon giving the bowl a slight turning movement, a wave will be set up that travels round the bowl until dissipated by friction. It will be seen that the line of the wave crest is a radius of the bowl, that the amplitude of the wave increases radially outwards, and that at the centre of the bowl the elevation of the water surface remains practically unchanged. The wave in the bowl is analogous to the tidal wave of a partly enclosed sea of roughly equidimensional plan. The line of the experimental wave-crest is, at any instant, a co-tidal line (line along which high water occurs at the same time) whilst a circle with centre at the centre of the bowl is a co-range line (line of equal tidal height). The central point of no change is an amphidromic (no tide) point. If neutrally buoyant particles (e.g. small pieces of coloured gelatin) are put in the water, it will be seen that the current beneath the rotating wave describes an orbit similar to that predicted for the real tide by Eq. (5.28). Such amphidromic systems as we have here are very common in real partly enclosed seas. In the North Sea, for example, there are two amphidromic points, one near the Norwegian coast and another, much more important, centrally placed. Amphidromic tides are also well known from the Black and Adriatic Seas and from the Persian Gulf. Even the tides of the open oceans seem to be amphidromic.

Tidal flow in real estuaries is strongly affected by viscous losses, and the assumption of frictionless conditions can lead to no more than an insight into the phenomena to be expected. Real estuaries are, moreover, channels that narrow and shoal away from the open sea, so that reflection of the tidal wave, leading to a standing wave, must also be considered. Many estuaries, too, receive a supply of fresh water from rivers that in some cases matches in volume the tidal prism. However, again assuming frictionless conditions, we can write for the tide in an infinitely long canal of uniform section,

$$\left.\begin{aligned} h &= A\cos(\eta t-\kappa x)\\ u &= \frac{A}{d}c\cos(\eta t-\kappa x)\\ c &= \sqrt{(gd)} \end{aligned}\right\}. \tag{5.29}$$

Now if the tide enters a canal of constant cross-section but closed at one end, complete reflection of the wave occurs and a standing wave is set up. If the maximum amplitude of the wave is A where it enters the canal, at the closed end the maximum amplitude is increased to $2A$ and occurs at multiple distances of $\lambda/2$. Points of zero amplitude, but maximum horizontal velocity, will be found at odd multiple distances of $\lambda/4$. The increase of tidal amplitude within a closed canal is consistent with the very large tidal ranges found in estuaries, for example, the Bay of Fundy and the Severn Estuary. The nodal points are only found in the very largest estuaries, however, since calculations using the equations of wave theory above will show that λ is considerable for the tide. Analyses can be made for frictionless tides entering canals of gradually diminishing cross-sectional dimensions, but we need here note only that the diminution of the cross-section has the effect of increasing tidal range and velocity. In real cases the increase proceeds to a limit set by the sources of energy loss.

5.4 Oceanic Circulations

The oceanic circulations depend on the fact that in equatorial regions of the earth, the heat energy received exceeds the energy lost, whilst in polar regions the loss exceeds the input. The resulting temperature differences between high and low latitudes require a poleward flow of heat from the equatorial regions. As conduction in the atmosphere and ocean is inadequate for the task, the heat is transferred chiefly by fluid motions. We can do no more here than very lightly sketch the main features of the resulting circulations, all of which are strongly affected by Coriolis force.

The surface winds associated with the major atmospheric circulations are sketched on the left-hand side of Fig 5.6. These winds exert a drag on the relatively warm, uppermost layer of the ocean lying above the thermocline (zone of rapid temperature and density change between warm relatively light water above and cool relatively dense water below) at depths between 50 and 900 metres. As the result of the drag, wind-driven circulations are set up in the uppermost layer, as sketched on the right-hand side of Fig 5.6 for an ideal ocean. It will be noticed that the predicted current is fastest and narrowest along the western borders of the ocean, as is found to be the case in real oceans. Actual current velocities measured for wind-driven circulations range between a few centimetres per second and a little over 2 metres per second.

The deeper oceanic circulations seem to depend on the sinking in the polar regions of water that has been cooled and thus made more dense. These currents occur below the thermocline and are associated with distinct water masses in a stratified series. The deep circulations are comparatively little known. In the Atlantic Ocean, however, there is a strong deep current running along the western continental border from Greenland to the Falkland Islands. In the northern Atlantic, the deep circulation apparently has an opposite rotation to the wind-driven surface current. The speed of the deep currents is ordinarily between about 1 and 10 cm/s, higher velocities being observed locally in areas of constricting, rough or steeply rising bottom.

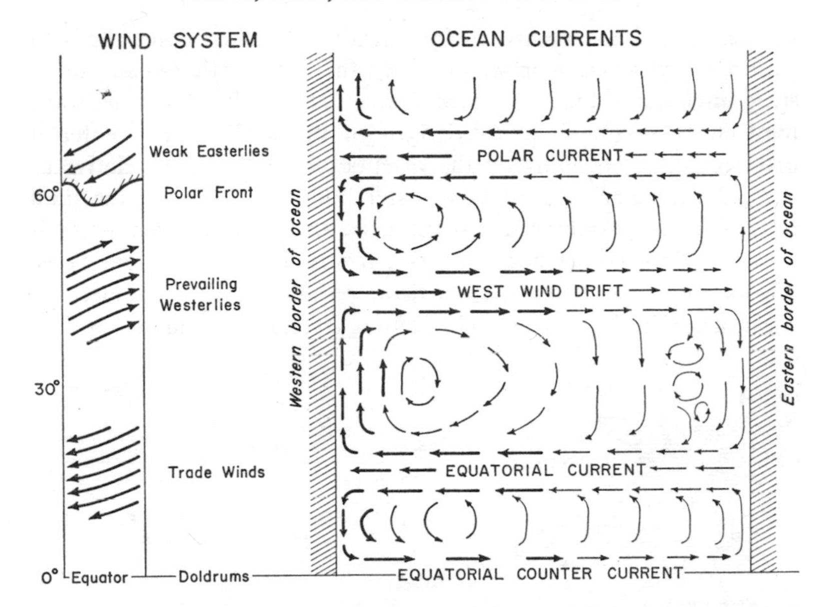

Fig 5.6 Near-surface ocean circulation in relation to global winds for an ideal ocean in the Northern Hemisphere (simplified from W. H. Munk). Thickness of arrows denotes strength of ocean current.

Both the wind-driven and deep currents are turbulent, though the intensity of the turbulence is generally small. A certain amount of turbulence is also generated during the travel of waves over the sea surface, but of course much more energy passes into turbulence inshore where waves break. The existence of turbulence in the sea means that marine currents have somc ability to transport fine sediment by turbulent diffusion. This process is particularly important near to land where rivers introduce mud.

5.5 Sedimentation on Open Shelves

One of the most important cases of shallow-marine sedimentation to consider is that on a broad, substantially rectilinear continental shelf, with an accretionary coastline supplied with mixed sediments, that borders an open ocean or large sea. Pertinent examples are numerous, being furnished by many deltas and coastal plains of alluviation (e.g. Nile and Niger deltas, Gulf of Mexico coastal plain).

In these cases, the waves chiefly affecting the shelf enter the area from the open ocean or sea, striking the coast orthogonally or at steep angles, and their spectrum changes seasonally. The tidal wave may either sweep along the length of the shelf or else enter it orthogonally, depending on the orientation of the shelf relative to the general shape of the whole water body. In any case, the tidal range is unlikely to exceed 1 metre, even at the coast, since the shelf borders a large water body and is unrestricted. In most real cases, a low-speed wind-driven current flows parallel to the coastline.

Since the shelf deepens with increasing distance from land (Fig 5.7),

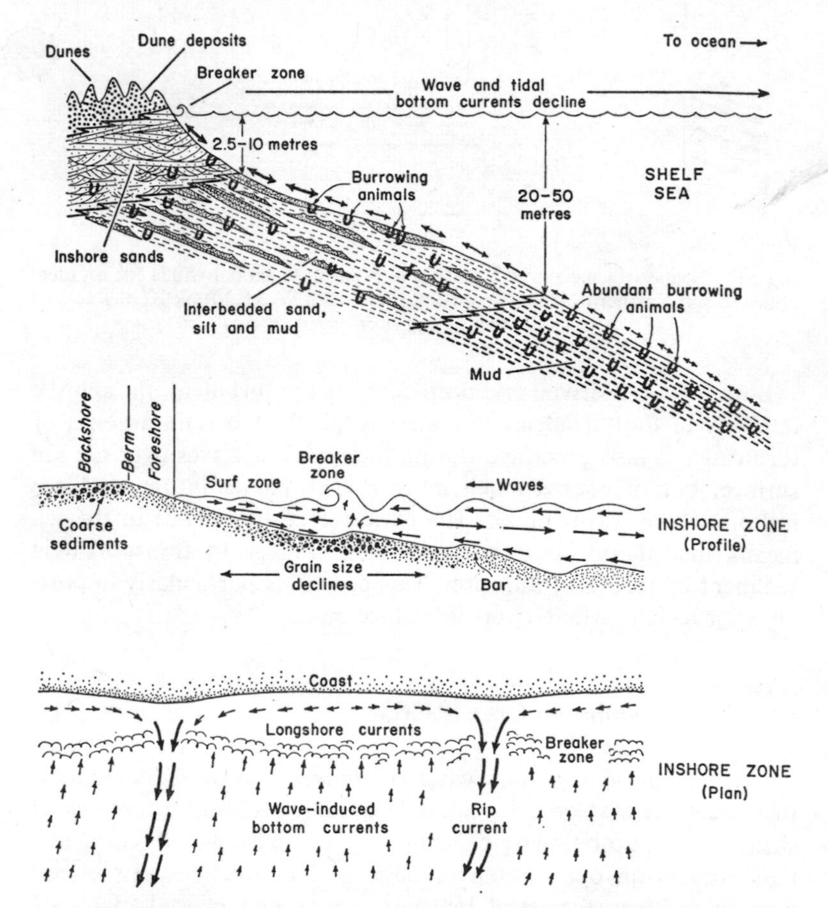

Fig 5.7 Model of sedimentation under wave and tidal conditions on a shelf exposed to the open ocean. Thickness of arrows denotes current strength.

it will be obvious from the preceding theory that the v
bottom water moving under the influence of wind wave
tide must gradually decrease between the coast and the s
Although the bed shear stress due to wind waves and the ti
be either accurately or simply expressed, we can suppose that the magnitude of the stress is directly, but not necessarily linearly, proportional to the water velocity near the bottom. Now sediment reaches the shelf at points on the coast, whence it is distributed parallel to shore by longshore and other currents. But the total system of currents is such that in addition sediment is moved outward from the coast, which may be considered a linear source. However, a particle of cohesionless sediment (coarse silt, sand, gravel) cannot travel into a depth greater than the depth at which the threshold stress for that size of grain equals the maximum instantaneous stress exerted on the bottom due to the simultaneous action of the various currents. Thus at any instant, for some defined wave spectrum and tidal range, we may imagine that there exists on the shelf an energy 'fence' which limits, in terms of distance from land and depth of water, the outward travel of particles of each size. For grains can be re-eroded only if they lie landward of the fence, where the maximum instantaneous stress exceeds the threshold. Of course, as we vary the wave spectrum and tidal range, the fence for a given size of particle changes somewhat in position, to define, over the full range of seasonal variation of conditions, a zone on the sea bed. Hence by applying these considerations to particles of a broad range of sizes within the general cohesionless class, we see that the bed-load sediment deposited on the shelf must decrease in grain size from the shore outward to deeper water.

The fate of silt and clay particles, fine enough to be carried by turbulent diffusion, can also be deduced, for we can assume that the magnitude of the turbulent fluctuating velocities increases with increasing velocity of the wave and tide-induced near-bed currents. Silt and clay are unlikely to be deposited where sand is being actively moved, as the turbulent velocities will be large enough to sustain mud in suspension. As deeper and deeper water is traversed, however, the magnitude of the fluctuations decreases, and there comes a depth when sand ceases to be moved and the turbulence is no longer capable of suspending the coarsest silt. As deeper water still is traversed, progressively finer mud becomes deposited. It is unlikely to be eroded, on account of the cohesive properties of the deposited material, unless the wave-induced currents due to a sudden storm drastically increase the bed shear stress. A belt of mud deposition will therefore be encountered on the shelf (Fig 5.7), though its position at any instant will vary with the properties of the tide and the waves.

The deposits of an open shelf therefore consist of an inshore belt of sand, with gravel on the beaches if this material is in supply, which grades out through progressively finer deposits into an offshore belt of mud (Fig 5.7). Because of seasonally variable waves and tides, a transitional zone between the sand and mud belts is to be expected. This inferred pattern is broadly matched in nature, though portions of it warrant closer inspection.

The physical environment is especially complicated near to shore. In the vertical, there is good field and experimental evidence for the circulation shown in Fig 5.7. This pattern is divided by the breaker zone into two pairs of cells. In plan (Fig 5.7) the near-shore circulation is dominated by mass transport currents, rip currents and longshore currents which also form cells. The total nearshore circulation strongly depends on the mass transport associated with the waves and on the wave motion and run-up in the surf zone.

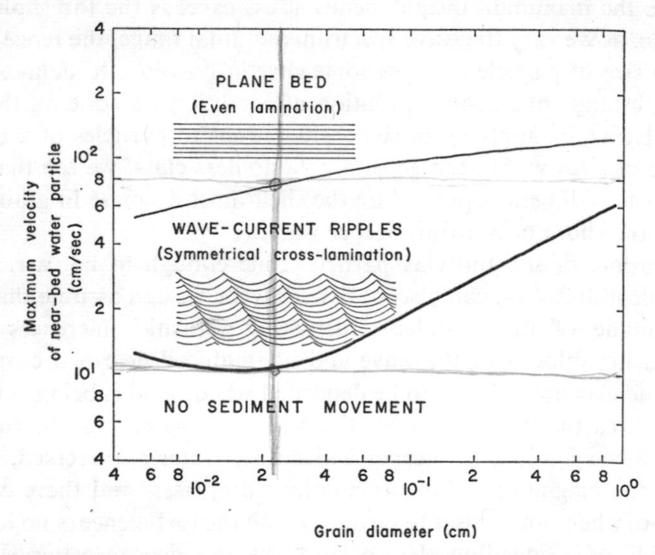

Fig 5.8 Bed forms in relation to maximum, wave-induced, near-bottom water particle velocity and calibre of quartz-density bed-material (data of M. Manohar and D. L. Inman).

Seaward of the breaker zone, the mass transport currents due to the waves shift outwards suspended particles, which include some sand, and move landward the near-bed grains that are disturbed. From Fig 5.8, which gives the threshold 'velocity near the bottom' for quartz sand grains, oscillatory currents of the order of 10 cm/s are seen to be all that are needed to set in motion the finest sand. Velocities of this magnitude can be induced by large waves at depths of 100–300 metres, and even the 'average' waves of open seas and oceans are capable of moving sand in depths of the order of 15–20 metres. The bottom sediments for some distance seaward of the breaker zone are found to consist wholly or partly of sand which coarsens landwards. Sand unaccompanied by mud is found to a depth of 10 metres on many open shelves and, interbedded with mud, it occurs to depths of up to 50 metres. The sands are moderately well to well sorted and are very commonly formed into symmetrical or near-symmetrical wave-current ripples, mostly trochoidal in profile. Fig 5.8 gives the range of near-bed velocities within which wave-current ripples are considered to form. Extensive field observations made on exposed shelves have shown that the wavelength of wave-current ripples is a function of grain size, and the orbital diameter and velocity of the wave currents, the theory of which was given above. For small orbital velocities the wavelength increases with orbital diameter, but for larger values decreases with orbital size (Fig 5.9). The largest wave-current ripples observed on open shelves are about 1 metre in wavelength.

The breaker zone is important in many respects. Here much of the wave energy is dissipated, some in turbulence and some in disturbing sediment. A portion of the energy is, however, carried by way of a transfer of mass into longshore currents landward of the breakers. The movement of sediment from both sides towards the breaker zone is now well attested by tracer studies and direct measurements of the velocity vector. There is a higher concentration of suspended sediment, mostly sand, in the zone than in any other part of the near-shore system, and the bottom sediment is coarser grained and more poorly sorted than elsewhere. Because of the high wave orbital velocities (Fig 5.8), ripples commonly die out some distance seaward of the breaker zone, to give place to smooth beds and hence an evenly laminated deposit. There is commonly a sand bar, with steep side facing land, beneath the breaker zone and quite often two or three similar but smaller bars evenly spaced to seaward. These

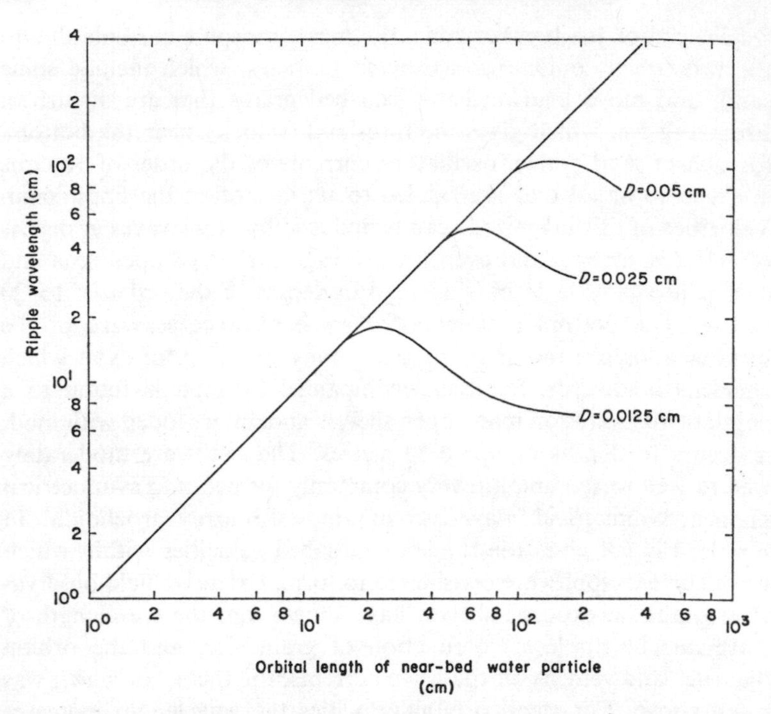

Fig 5.9 Wavelength of wave-current ripples in relation to length of trajectory of near-bed water particles and calibre of quartz-density bed-material (data of D. L. Inman).

bars all change their position and shape with variation in the wave condition. On many open coasts, the breaker zone is breached at fairly regularly spaced points by rip currents formed from the wave-driven water piled against the shore in the surf zone. These currents attain velocities sometimes exceeding 100 cm/s, and in places have gouged out shallow channels at steep angles to the shore, whose beds carry seaward-facing dunes.

Landward of the breaker zone is the surf zone, which extends up on to the foreshore of the beach, and whose highest point is the storm-season berm. The principal currents of the surf zone, forming a double circulation (Fig 5.7), are the longshore currents and the surge, running up the beach as swash and down it as backwash, that results from wave-breaking. The foreshore varies greatly in character, partly in response to the grain size of the beach material, and partly due to wave height and tidal range.

Gravel beaches are steep, even when wave activity is small, and under extreme conditions stand at a maximum angle equal to the angle of initial yield of the beach material. Beach gravels are invariably well sorted and rounded, the clasts showing abundant percussion marks, though at depth in a beach there may be a sand fill. Repeated working over by waves brings about a marked shape-sorting in beach gravels.

Of greater importance, however, are sand beaches. Where the tidal range is narrow, the beach may be no more than a few tens of metres wide. Very commonly, its surface is smooth and underlain by evenly laminated, well sorted sand, for the swash and backwash flows under these conditions attain the power of the plane bed regime (Fig 2.6). The steepening of the beach during the storm season of taller waves is reflected in the beach by erosion surfaces between successive sets of evenly laminated sand. Shallow seaward-facing troughs representing buried beach cusps may also be found internally. A large tidal range is commonly associated with a beach several hundred metres wide diversified into bars and troughs lying nearly parallel to the coast. Complex and changeable patterns of water and sediment movement are generated as the sea moves back and forth over such beaches. The seaward faces of the beach bars are generally smooth and flat, with parting lineation, but the landward sides are very often at the angle of repose, the bars presumably being cross-bedded internally. In the troughs between the bars, current ripples and dunes arise during the tidal ebb. Wave-current ripples form in sheltered areas where the water undergoes a degree of ponding. Hence many different types of cross-stratification are found in foreshore deposits. Foreshore sediments are generally well sorted, and finer grained than the sediments of the backshore.

Few data exist on physical conditions in the area beyond the sand zone in our model (Fig 5.7), but, though the energy density of this part of the general environment is comparatively low, there are no reasons for regarding conditions as constant. Indeed, the existence of a transitional zone of layered mud and sand is itself an indication of considerable variation. Here sand and coarse silt in layers are interbedded with silty clays and clayey silts. The coarser layers are thickest at the landward edge of the zone, and thinnest and least numerous at its seaward margin. Wave-current ripples, and sometimes even laminations, are found in the coarse layers. These layers, with sharp and in places erosional bases, are very commonly graded from coarse up to fine, suggesting that the local energy of the environment at first increased with the influx of sand and then decreased as the sand ceased to move and mud deposition returned. However, the energy changes are unlikely to have been the work of minutes. Storm conditions are the most likely cause of the repeated spread of sand and silt over mud.

Above the mud belt (Fig 5.7) the bottom water is unsufficiently turbulent to keep suspended even those particles whose free falling velocity is of the order of 0·01 cm/s. Deposition occurs floccule by floccule, or actually flake by flake, but is probably not continuous, as exceptionally large waves could lead at the bottom to non-deposition or slight erosion. Lamination may be present but difficult to detect.

Open shelf sediments, particularly of the transitional and mud belts, are subject to extensive reworking by bottom-dwelling and feeding animals, even to the extent that all traces of primary laminations are destroyed.

5.6 Sedimentation in Shallow Partly Enclosed Seas

The sedimentary environment of shallow, partly enclosed seas is determined chiefly by the strength and direction of tidal currents, whereas on open shelves, waves are predominant. There are many examples of tidal sedimentation in shallow seas, but only the situation in waters around Britain is at all well known. The work here has been concentrated on the southern North Sea, the English Channel, and the Celtic Sea (Fig 5.10). The North Sea is essentially a rectangular basin open along one of its short sides, while the Celtic with the Irish Sea and the English Channel form two nearly mutually perpendicular gulfs. The debris involved in recent sedimentation in this area has several origins. We find that much of the sand and gravel was introduced during the Pleistocene as till or fluvial sediment, and has subsequently been reworked during and following on the later sea-level rise. A smaller part was introduced by rivers during the rise, another smaller part was got by the marine erosion of the rock substrate and surrounds, and a third, smaller part is biogenic. Some of the mud represents reworking, though probably most was introduced by the major rivers.

The region is comparatively shallow and much is known of its

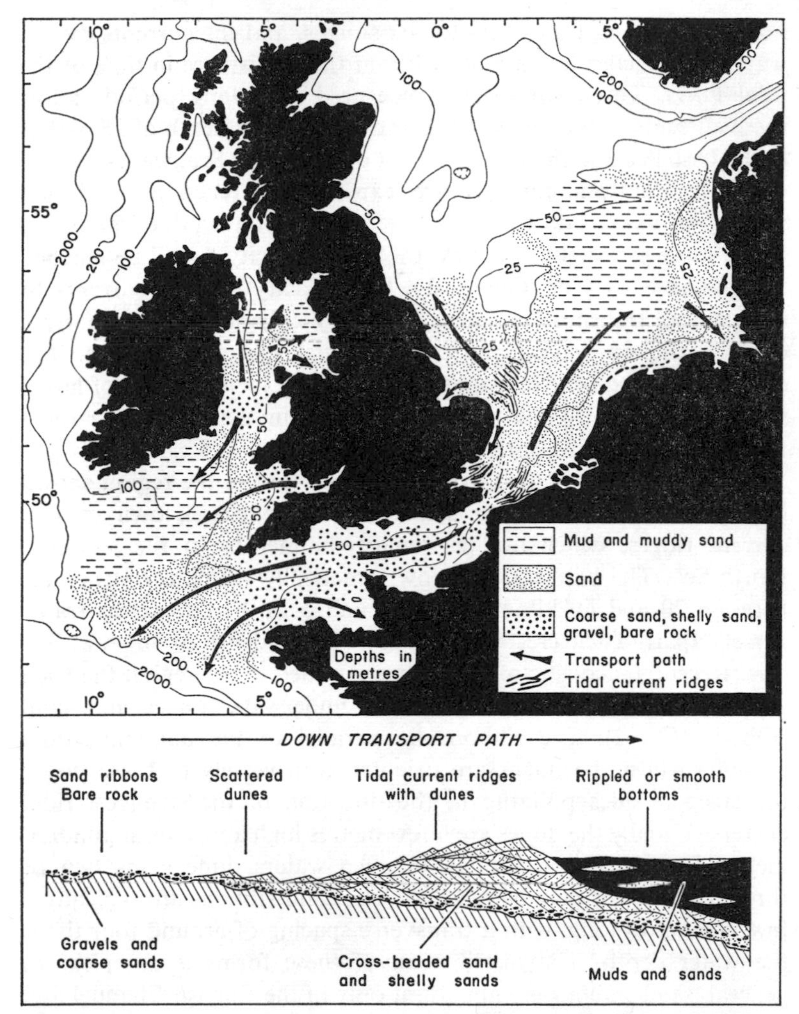

Fig 5.10 Features of sedimentation in the seas surrounding the British Isles (compiled from data of R. H. Beldersen, G. Boillot, J. J. H. C. Houbolt, A. H. Stride and British Admiralty Hydrographic Department). The inset profile shows the changes typically encountered downcurrent along a transport path.

waves and tidal currents (Fig 5.10). There are large areas, mostly away from land, where the maximum tidal currents reach velocities between 50 and 100 cm/s, and smaller areas, chiefly close to land or to shoals, where these currents range between 100 and 200 cm/s. Locally the maximum tidal currents exceed 200 cm/s. The area is notorious amongst mariners for its storms, and the currents due to waves are spectacular, though without the persistence in time of the tidal flows. The wave spectra measured over long periods allow these currents to be calculated. For example, at a depth of 25 metres in the Irish Sea and the southern part of the North Sea, wave-induced oscillatory bottom currents are expected to reach a maximum velocity of 30 cm/s on 10 days in every year. At a depth of 50 metres in the more exposed Celtic Sea, the same velocity would be reached 30 days in each year. Obviously, the capacity of tidal currents to disturb and transport sediment is greatly enhanced by the bottom currents due to waves, particularly under storm conditions. The complexity of the currents makes difficult the identification of large-scale circulations in the region. However, in the North Sea there appears to be a slow anticlockwise motion.

It has long been known from soundings that in the region shown in Fig 5.10 there exist large sand or gravel banks, known as tidal current ridges, which are particularly numerous in the southern North Sea. The banks are narrow, practically rectilinear structures between 20 and 60 kilometres long and 5 or 10 kilometres transversely apart. Their crests lie at a few metres depth or shoal at low tide. Recently, acoustic mapping has revealed in the region the wide occurrence of two other bed forms, underwater dunes and sand ribbons (Ch. 2). The ribbons lie parallel to the dominant tidal currents whilst the dunes are oriented transversely to these flows, the steep leesides pointing in the direction of the strongest tidal current. Usually the dunes are a few metres high and several hundred metres in wavelength, though in deep waters dimensions two or three times larger may be reached. The sand ribbons are typically a few kilometres long, with a transverse spacing of around four times the water depth. The distribution of these forms is complex. In general, sand ribbons are abundant only in the English Channel and the inner Celtic Sea. Dunes are most abundant in the southern part of the North Sea, but are encountered patchily in the Celtic and Irish Seas.

From the orientation of bed forms of these three kinds, it is clear that coarse sediment is moved *parallel* to land by the tidal currents, with some movement up the larger estuaries (Fig 5.10). The pattern of movement is marked by nodes and foci amongst the transport paths, which are individually rather short. As each transport path is traced out, important changes in the sea bed are found.

The heads of the transport paths are characterized by the erosion of the substrate, which is locally bared, into streamwise trenches and grooves several metres deep. Dunes and sand ribbons scattered in the vicinity testify to the movement of material away from the head of the path. Downstream, a bed of gravel, commonly rich in mollusc, bryozoan and ophiuroid debris, is found resting disconformably on the eroded substrate. Such gravels occur extensively in the English Channel, especially around the Cherbourg Peninsula. The gravel beds, probably thin, are themselves surmounted by sand ribbons just a few centimetres thick. The sand ribbon zone leads downstream to an area in which the basal gravel is completely buried under a thick layer of sand whose surface is fashioned into dunes, often carried on tidal current ridges. The sands are known from grab samples to be chiefly fine or medium grained and locally gravelly or shelly. The shells are broken, abraded and stained, and clearly have been much reworked. Coring has shown that the dunes found in these areas are internally cross-bedded. Geophysical explorations have revealed that the tidal current ridges themselves show internally a low-angle bedding pointing to their gradual movement. A number of the transport paths end in a region where mud and very fine sand have been deposited to a thickness of, in places, 30 metres. The most important mud areas occur in the central North Sea and south of Ireland.

5.7 Sedimentation in Estuaries and Tidal Flats

An estuary is a funnel-shaped area of tidal sedimentation connected along its broad base to the open sea and, commonly, at its narrow apex to a river. A tidal flat, also an area primarily of tidal sedimentation, is flask-shaped in plan or consists of a series of flask-shaped sub-areas in lateral contact. Each flask-shaped area or sub-area is a tidal drainage basin analogous to the drainage basin of a river on land, and each drainage basin is connected to the open sea by a narrow channel, represented by the neck of the flask, lying between the tips of sandy barrier islands. Because of these shape differences, the tidal currents of a tidal-flat area are strongest in and near the inlet channels, and weakest in the peripheral zones far from the inlets. Wave action is comparatively low on account of the protective barriers. In an estuary, on the other hand, the tidal currents tend to be spatially uniform in strength, whilst wave action becomes comparatively significant. The vertical and lateral sequences of sediments in tidal flats and estuaries are much the same, but the differences of shape and current activity lead to a different facies geometry and balance in the one environment than in the other.

Fig 5.11 shows the distribution of major sedimentary environments in a fictional estuary with a river at the head. As we saw in the preceding section, most of the sediment can be regarded as introduced into the estuary from the sea, the river supplying a small proportion only. The mean level of low tides serves as a critical height, for it separates that part of the estuary affected at practically all times by wave and tidal currents from that part which is flooded only a part of the time and otherwise exposed to the air. In the sub-tidal depths

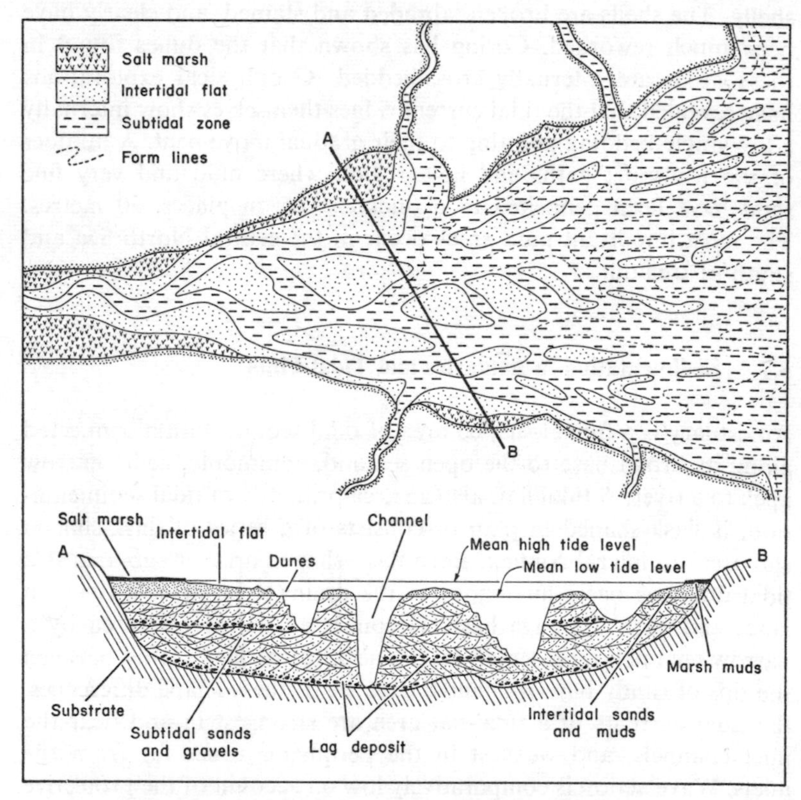

Fig 5.11 Model of sedimentation in an idealized estuary.

below mean low tide level we find a complex of sediment banks which define interconnected as well as branched channels. The banks in plan vary from long and narrow, or compressed sigmoidal, in the lower part of the estuary, to oval or triangular in the upper reaches. Some banks in the upper reaches are attached to the shore, the channels between them showing an obvious meandering to braided pattern. As the estuary is ascended, the area above mean low tide level increases relative to the subtidal area. The zone lying between mean low tide and mean high tide levels form the inter-tidal flat, commonly divided into a lower and an upper part on the basis of degree of exposure to tidal current action. Above mean high tide level are the salt marshes inundated only during the highest tides. It will be noticed that the relative proportion of inter-tidal flat increases headwards in the estuary, and that the salt marsh is restricted to sheltered marginal areas.

The same three main environments exist in tidal-flat areas but in a different spatial distribution and relative abundance (Fig 5.12). The sub-tidal area is comparatively small and, except in the lower reaches of each drainage system, the channels are meandering and dendritic. The inter-tidal flat is broad and encircling, carrying the smaller channels of the dendritic drainage system, which are dry at most stages of the tide. Beyond the inter-tidal flat are extensive peripheral salt marshes crossed here and there by the tidal channels of streams draining into the flat.

Estuarine and tidal-flat sediments are best known from Europe and North America. The most extensive work has been done on the flats and estuaries bordering the North Sea and the Atlantic Coastal Plain of the USA, and the following account will refer mainly to these areas. Of tidal-flat and estuarine sediments in warm or hot climates little is known, except that the larger and more luxuriant halophytic plants of these regions introduce into the geomorphology and sediments some special features.

The deposits of the sub-tidal channels and shoals of estuaries and tidal flats and chiefly well sorted fine to medium grained sands, shelly sands, and gravelly sands which rest disconformably on an irregular and often channelled surface cut into older deposits below. At estuary mouths, and in the deeps of the inlet channels of tidal flats, the substrate may actually be exposed over considerable areas of the bottom, or covered by only a thin lag gravel. The shoals rise between 10 and 30 metres above the floors of the adjacent channels

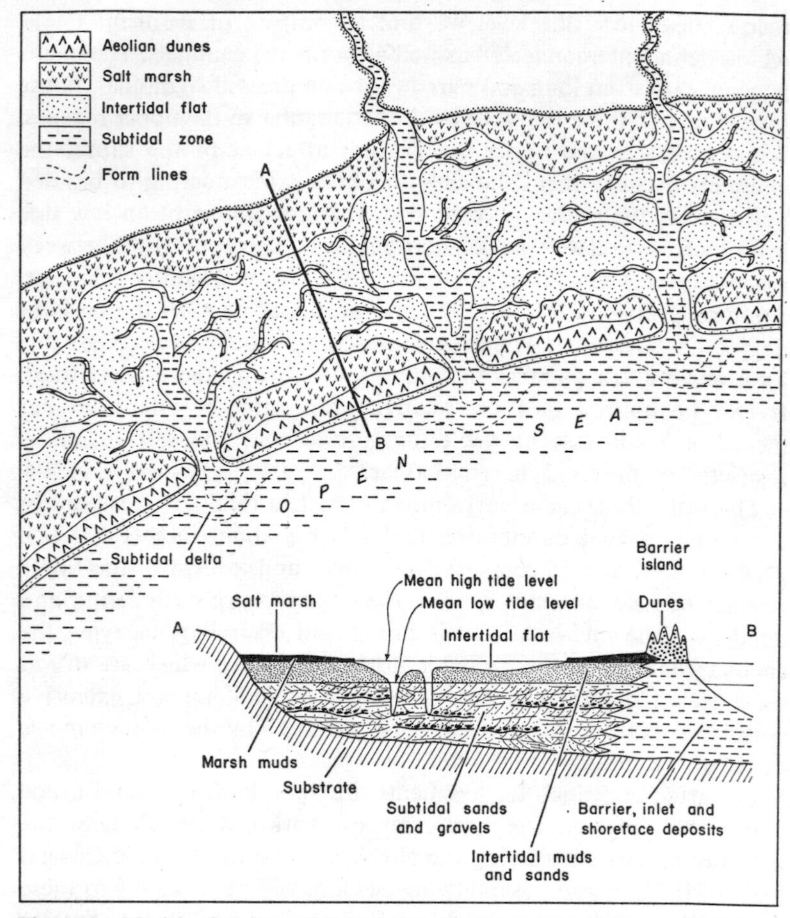

Fig 5.12 Model of sedimentation in an idealized tidal flat protected by a barrier island chain.

and have some degree of mobility under the changing current regime. There is normally within the channel complex a segregation of the flood and ebb currents along different paths, as can be inferred from the sigmoidal form of many shoals. The sides of the shoals and the channel floors bear dunes ranging in height between several decimetres and a few metres. Internally, the shoals are mainly cross-bedded, though cross-laminations due to current ripples, as well as even laminations, are locally abundant. Thin layers of mud are sometimes found in the deposit and presumably denote unusually calm conditions or places of shelter. Many impersistent shell or gravel layers occur in the shoals, a token of prolonged if local erosion. Because of the ebb and flow of the tide, combined with the mobility of the banks, the internal structures denote a wide range of current direction. A herring-bone arrangement between the cross-strata of successive units is far from rare.

The inter-tidal flat can in many cases be divided into a lower zone in which sand is the main deposit, and an upper one in which the sediment is predominantly mud.

The lower sand flats consist of well sorted sands, normally fine and very fine grained, which represent an upward extension, into the inter-tidal zone, of the kind of deposition represented by the sub-tidal shoals. Commonly the sand flats reveal small dunes that can be traced downward into the permanently concealed dunes of the channels. Symmetrical wave-current and asymmetrical current ripples, the latter varying from straight to linguoid in plan, are also often found on the sand flats. Where the flats are exposed to comparatively strong wave action, mainly at the windward edges of the channels, a smooth sand surface underlain by even laminations is encountered. Mud is on occasions deposited on the lower inter-tidal flats, principally in the troughs of the dunes and ripples, in sheltered scour hollows, and in the lee of the larger undulations on the surface of the flats. On the drowning of the surface and the renewed movement of the sand, some of the mud deposit becomes broken down into lumps or flakes which are incorporated as gravel into the sand. These mud gravels are frequently seen amongst the foreset deposits of the dunes. Faunally, the lower sand flats are poor in number of species but uncommonly rich in individuals of one or two species. Biogenetic structures abound in the lower inter-tidal flat deposits. In Europe, the lug-worm *Arenicola* is very commonly the chief burrower in the lower flats. The decapod crustacean *Callianassa* takes the place of *Arenicola* in the warmer waters of the Atlantic Coastal Plain.

In the upper inter-tidal flats, mud predominates because the area is inundated only for a few hours ranging around the time of high water, when the tidal currents are of greatly diminished strength. Usually, the deposit found in the lower part of the upper flats consists of layers a few millimetres or centimetres thick of sandy mud and sand. The sand layers ordinarily show wave-current or current ripples, but may in places be evenly laminated. Isolated sand ripples buried in mud are not uncommon. Upwards on the flats the sand layers tend to become thinner and less numerous, the deposit being almost wholly sandy mud. At times of neap tide, the muds of the upper flats become sun-dried and deeply cracked, their later erosion being thereby facilitated. Shallow linear scours a few centimetres or decimetres deep, and locally infilled with mud flakes and shells, are common, as are flatter erosional surfaces of comparatively wide extent from which project the empty but still articulated shells of once-concealed burrowers. The muds of the upper flats provide a rather favourable environment for burrowing worms, molluscs, gastropods and crustaceans. Often the density of the burrows is considerable, leading to the wholesale destruction of stratification in the sediments.

The inter-tidal flats are traversed by broad, shallow gullies and runnels empty save at high tide. These channels are meandering and they vary their position in the same way as the curved channels of rivers. On the inner bank of each curve a tidal point bar, or lateral deposit, is formed, which rests on a lag gravel of shells and lumps of mud accumulated on the channel floor after the erosion of the outer bank. On the upper part of the flats, where the gullies and runnels are most numerous, the lateral deposits consist of alternating layers of sand and muddy sand.

The halophytic vegetation of tropical and subtropical lands includes such trees as the mangrove. These vigorous and aggressive plants are capable of spreading out over what in temperate climates corresponds to the upper part of the inter-tidal flat. The mangrove builds a dense root-mat within the tidal mud and, by virtue of its prop-roots, greatly encourages mud deposition. Moreover, the spreading roots and rootlets bind the sediment and enhance its resistance to erosion. Mangrove-covered inter-tidal flats are well known from West and East Africa, the East Indies and parts of Australia, and tropical South America. One of the features of these regions is that the tidal channels of the flats are reticulate. This pattern, so different from the dendritic pattern seen on the bare flats of temperate regions, appears attributable to the role played by the mangroves in promoting deposition and in binding the deposit.

Deposition in the salt marshes of tidal flats and estuaries is much affected by atmospheric agencies and biological activities. The sediments are chiefly very fine sand, muddy sand and mud in irregular, rapidly alternating layers, though in many marshes the deposit is almost wholly mud. There may also be layers of shells thrown up during storms from the lower lying inter-tidal flat. Another important constituent is organic debris resulting from the attrition and decay of marsh plants. The occurrence of these plants is controlled partly by their salt tolerance and partly by the general muddiness of the sediments. The rice-grass *Spartina* is an especially important marsh plant in North Sea regions and in the Atlantic Coastal Plain. One of the most characteristic features of salt marshes are the shallow, irregular hollows called salt pans. Mud settles in these pans when the marshes are flooded, but afterwards dries out and cracks. Whilst briefly flooded, the pans support communities of algae and brackish-water animals. Marsh sediments are generally much burrowed and fairly well oxygenated.

5.8 Ancient Shallow-Marine Deposits

The recognition of ancient shallow-marine deposits depends chiefly on finding salinity indicators, the most important being distinctive floras or faunas (trace as well as body fossils), evaporite beds that could have been precipitated from sea water, and horizons of halite pseudomorphs. Mechanical sedimentary structures and lithology, at least as represented in terrigenous clastic deposits, are of little significance as salinity indicators. They are, however, when combined with the overall geometry of the deposit and its main facies, of great value in deciding which of the shallow-marine models is most representative of a particular case.

At several periods in the history of the earth there has been extensive transgression of the sea over nearly flat platform or shield areas. Several deposits resulting from such transgressions find parallels in the sediments formed in present-day partly enclosed tidal seas, either because the ancient seas were themselves partly enclosed or because, being extensive but of relatively uniform depth, strong currents were developed everywhere in them. Examples are afforded by Cambrian and Cretaceous rocks.

South of Lake Superior, the Upper Cambrian includes the Munising Formation, developed over an area measuring 260 by 80 kilometres. The Munising is a shallow-marine deposit, no more than 80 metres thick, that begins with a conglomerate, attaining a maximum thickness of 5 metres, composed of very well sorted and rounded pebbles. Above the conglomerate are fine and medium grained moderately well to well sorted sandstones showing chiefly trough cross-bedding probably due to dunes and, at limited horizons, sun cracks and wave-current and asymmetrical ripple marks. Clay flake conglomerates and trace fossils occur sporadically. Shales are common only in the upper part of the Munising.

A second formation resembling the deposits of a partly enclosed tidal sea is the Lower Cambrian quartzites of the Northwest Highlands of Scotland. This formation, only 150 metres thick at the greatest but extending laterally over an outcrop of 160 kilometres, has a pebbly bed at the base followed by fine to medium grained, well-sorted quartzitic sandstones that are evenly laminated or cross-bedded, and in places ripple marked. At certain levels, notably in the Pipe Rock, there are burrowings at a density of many hundreds per square metre. The marine fauna is, however, first preserved only towards the top of the quartzite sequence.

The Cretaceous Folkestone Beds and Woburn Sands of southern and south-eastern England can without much doubt be attributed to deposition in a partly enclosed tidal sea. These formations, which rarely exceed 100 metres in thickness but have an outcrop 380 kilometres in length, are chiefly fine to very coarse grained well sorted cross-bedded sands with a sparse ammonite fauna. The cross-bedded units vary in thickness from a few decimetres to as much as 5 metres, and presumably represent large dunes. In places, the cross-bedded units fill channels cut in older sands of the formation. Analysis of the cross-bedding shows that the currents flowed from north-west to south-east, parallel to the ancient sea-shore. At many localities, thin layers of mudstone are seen draped over the foresets and bottomsets of the cross-bedded units, suggesting that in places on the sea floor the dunes were active only during spring tides or at times of storm-enhanced currents. The sands are intensely burrowed and contain much chiefly oxidized glauconite.

It must surely be the case that rocks of estuarine or tidal flat origin exist in the geological record, though a convincing study of such beds has yet to be published. A tidal flat origin has, however, been claimed for Upper Devonian sandstones in Belgium and Germany and for Middle Jurassic shales and bioclastic limestones in the English Midlands.

Many studies have been made, particularly by petroleum geologists, of shallow-marine deposits that accumulated on open shelves as a part of a spreading deltaic or coastal plain complex.

An important example is afforded by the Devonian Baggy Beds of south-west England. This formation is 440 metres thick, but of unknown lateral extent, and consists of several distinct shallow-marine facies, many rich in trace fossils and some in invertebrate remains. The bulk of the sediments comprise shales, siltstones and very fine sandstones with structures, such as wave-current ripples, indicating their affinity with the transitional layered deposits of the model (Fig 5.7). The more important of the other facies can be attributed to tidal channels, dune-covered sand flats, or sandy areas extending some way offshore from the beach. However, it is not certain whether the Baggy Beds form part of a deltaic or coastal plain complex, as their lateral variation is insufficiently known. A similar association of sediments to the Baggy Beds has been recognized and described from the Koblenz Quartzite of the Rhineland.

The Cretaceous seas which occupied vast areas of the western interior of the United States were broad and shallow, and were the site of accumulation of interfingering sandstone and shale bodies which represent the repeated advance and retreat of the shoreline of a broad coastal plain of alluviation. Much is known of these rocks, as they are relatively flat lying and widely exposed in a dissected terrain, and have locally been explored by boring. The marine rocks are shales and siltstones which pass upwards and laterally into sandstones. In favourable outcrops, even the profile of the ancient beach face and sea bed can be made out in the shape of major bedding surfaces. The fauna, of body and trace fossils, is distinctive and includes many thick-shelled bivalves. The sandstones are of two main kinds, sheet and channel. The sheet sandstones appear from their sedimentary structures to be beach and subtidal deposits, but the mainly cross-bedded channel sandstones seem to vary between fresh water and tidal. Locally, thick coals accumulated in low-lying marshes back from the shore.

READINGS FOR CHAPTER 5

The subject of oceanic circulations, waves and tides is large and complicated, and probably no single treatment can be regarded as entirely satisfactory. However, a valuable and readable account, lacking only in the more recent theories of wave generation, is given by:

DEFANT, A. 1961. *Physical Oceanography*. Pergamon Press, Oxford. 2 vols;

whilst a popular discussion of waves and tides is to be found in:

TRICKER, R. A. R. 1964. *Bores, Breakers, Waves and Wakes*. Mills and Boon Ltd., London, 250 pp.

An excellent sketch of the geomorphology of coasts and shallow seas is published by:

GUILCHER, A. 1958. *Coastal and Submarine Morphology*. Methuen and Co., London, 274 pp.

Some account of sediments being formed at present on open shelves is to be found in:

ALLEN, J. R. L. 1965. 'Late Quaternary Niger delta, and adjacent areas: sedimentary environments and lithofacies.' *Bull. Am. Ass. Petrol. Geol.*, **49**, 547–600.

BERNARD, H. A. and LEBLANC, R. J. 1965. 'Résumé of the Quaternary geology of the Northwestern Gulf of Mexico Province.' *The Quaternary of the United States*, Princeton University Press, pp. 137–185.

NOTA, D. J. G. 1958. 'Sediments of the western Guiana Shelf.' *Meded. LandbHoogesch. Wageningen*, **58**, 98 pp.

Aspects of sedimentation in the shallow seas around the British Isles are discussed by:

BELDERSON, R. H. and STRIDE, A. H. 1966. 'Tidal current fashioning of a basal bed.' *Marine Geol.*, **4**, 237–257.

BOILLOT, G. 1964. 'Géologie de la Manche Occidentale.' *Annls. Inst. Océanogr., Monaco*, **42**, 220 pp.

HOUBOLT, J. J. H. C. 1968. 'Recent sediments in the Southern Bight of the North Sea.' *Geologie Mijnb.*, **47**, 245–273.

OFF, T. 1963. 'Rhythmic linear sand bodies caused by tidal currents.' *Bull. Am. Ass. Petrol. Geol.*, **47**, 324–341.

ROBINSON, A. H. W. 1960. 'Ebb-flood channels systems in sandy bays and estuaries.' *Geography*, **45**, 183–199.

STRIDE, A. H. 1963. 'Current-swept sea floors near the southern half of Great Britain.' *Q. Jl. geol. Soc. Lond.*, **119**, 175–199.

The following accounts of the geomorphology and sediments of estuaries and tidal-flat areas are particularly useful:

BAJARD, J. 1966. 'Figures et structures sédimentaires dans la zone intertidale de la partie orientale de la Baie du Mont-Saint-Michel.' *Revue Géogr. phys. Géol. dyn.*, **8**, 39–111.

EVANS, G. 1965. 'Intertidal flat sediments and their environments in the Wash.' *Q. Jl. geol. Soc. Lond.*, **121**, 209–245.

KLEIN, G. DEV. 1963. 'Bay of Fundy intertidal zone sediments.' *J. sedim. Petrol.*, **33**, 844–854.

LAND, L. S. and HOYT, J. H. 1966. 'Sedimentation in a meandering estuary.' *Sedimentology*, **6**, 191–207.

LAUFF, G. H. 1967. *Estuaries*. American Association for the Advancement of Science, Washington, D.C., 757 pp.

REINECK, H.-E. 1963. 'Sedimentgefüge in Bereich der südlichen Nordsee.' *Abh. senckenb. naturforsch. Ges.*, **505**, 138 pp.

Environmental studies of shallow-marine sediments are numerous, but amongst the more interesting from the viewpoint of the present treatment are:

GOLDRING, R. 1967. 'The shallow marine and deltaic facies of the Baggy Beds (Upper Devonian) at Baggy, North Devon.' *Proc. geol. Soc. Lond. No.* 1643, 241.

HAMBLIN, W. K. 1958. 'Cambrian sandstones of Northern Michigan.' *Publs Mich. geol. biol. Surv.*, **51**, 146 p.

LANE, D. W. 1963. 'Sedimentary environments in Cretaceous Dakota Sandstone in northwestern Colorado.' *Bull. Am. Ass. Petrol. Geol.*, **47**, 229–256.

MACKENZIE, D. B. 1965. 'Depositional environments of Muddy Sandstone, western Denver Basin, Colorado.' *Bull. Am. Ass. Petrol. Geol.*, **49**, 186–206.

NARAYAN, J. 1963. 'Cross-stratification and palaeogeography of the Lower Greensand of south-east England and Bas-Boulonnais, France.' *Nature, Lond.*, **199**, 1246–1247.

NIEHOFF, W. 1958. 'Die primär gerichteten Sedimentstrukturen, insbesondere die Schrägschichtung im Koblenz quarzit am Mittelrhein.' *Geol. Rdsch.*, **47**, 252–321.

PEACH, B. N., HORNE, J., GUNN, W., CLOUGH, C. T. and HINXMAN, L. W. 1907. 'The Geological Structure of the North-West Highlands of Scotland.' *Mem. geol. Surv. U.K.*, 688 pp.

SABINS, F. F. 1964. 'Symmetry, stratigraphy and petrography of cyclic Cretaceous deposits in San Juan Basin.' *Bull. Am. Ass. Petrol. Geol.*, **48**, 292–316.

SHELTON, J. W. 1965. 'Trend and genesis of lowermost sandstone unit of Eagle Sandstone at Billings, Montana.' *Bull. Am. Ass. Petrol. Geol.*, **49**, 1385–1397.

YOUNG, R. G. 1955. 'Sedimentary facies and intertonguing in the Upper Cretaceous of the Book Cliffs, Utah-Colorado.' *Bull. geol. Soc. Am.*, **66**, 177–201.

YOUNG, R. G. 1960. 'Dakota Group of Colorado Plateau.' *Bull. Am. Ass. Petrol. Geol.*, **44**, 156–194.

CHAPTER 6

Turbidity Currents and Turbidites

6.1 General

A turbidity current is a species of gravity current, and in order to understand the former it is necessary to have some knowledge of the latter.

A gravity current is an intrusion of one fluid over, into or under another fluid generally of larger extent, the current arising because of a difference of hydrostatic pressure-force between the two fluids. The difference of pressure-force in turn depends on a dissimilarity of specific weight between the fluids in relative motion. In the case of intrusion below the ambient fluid, a stream of heavy fluid flows along the bottom and displaces the lighter medium. If in this case the bottom slopes in the direction of the current flow, then the body force is called into play to drive the current along, in addition to the hydrostatic pressure-force. When intrusion takes place along the free surface of the ambient fluid, the displacing stream consists of the lighter fluid, and the flow is in appearance the mirror-image of the previous example. In either case the stream of fluid maintains a characteristic longitudinal profile (Fig 6.1). There is a spatulate head wave which, independently of the scale of the current, rises to about twice the mean height of the interface of the stream following behind the head. On the rearward side of the head wave is a strongly disturbed zone indicative of turbulent mixing of the gravity current into the medium.

In nature, gravity currents are varied and frequent. They occur on a grand scale in the atmosphere as cold fronts and sea-breeze fronts and, on a smaller scale, are found in volcanic districts as *nuées ardentes* and in mountainous areas as avalanches of light snow or sometimes rock. Gravity currents occur in estuaries as intrusions of salt water. Where a river empties into the open sea, the fresh water commonly spreads out over the salt as a gravity current at whose edge foam and flotsam abound. However, mud-laden river water entering a fresh-water lake or reservoir is often so heavy that it plunges down to form a turbid underflow capable of travelling many kilometres over the bottom.

Turbidity currents are a class of gravity current with an excess of specific weight arising from the sediment suspended in them. They can occur where a river enters a body of fresh water, as we have just seen, but are probably most often produced by a transformation of underwater sediment slumps generated in regions of active deposition and moderate to steep slope. The geological importance of turbidity currents lies in their apparent ability to shift coarse clastic or bioclastic sediment in large quantities from shallow to deep water, especially in the seas and oceans, and thus to modify underwater topography and fill up deep sedimentary basins where a slow, hemipelagic style of deposition is the rule. The deposits of turbidity currents, or turbidites, are chiefly marine. Lacustrine turbidites are also known but are less common and important volumetrically.

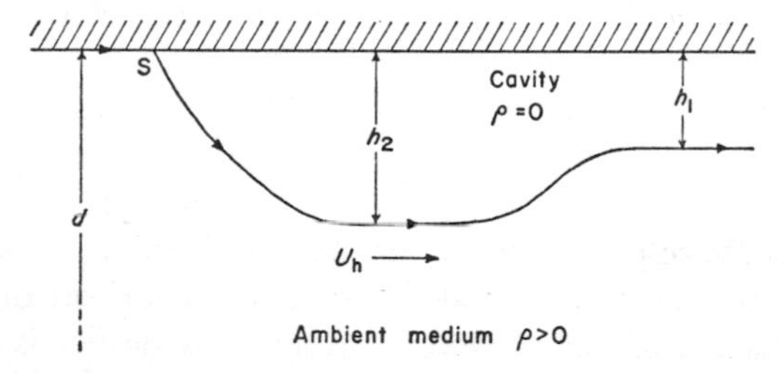

Fig 6.1 Definition diagram for theory of gravity current flow.

6.2 Motion of the Head of a Gravity Current

In order to make a simple mathematical model of the headward part of a gravity current, we suppose that an empty cavity is steadily advancing just beneath the upper solid boundary of an infinitely deep body of fluid of density ρ (Fig 6.1). We arrest the current by assigning to the ambient medium a velocity U_h equal and opposite to the velocity of advance of the cavity, and ignore effects arising from surface tension and viscosity. Since S is a stagnation point, the Bernoulli theorem states that the pressure in the cavity relative to that on the horizontal boundary far upstream is $p = \frac{1}{2}\rho U_h^2$. But the pressure in the cavity is also equal to the hydrostatic pressure $p = \rho g h_1$, where h_1 is the net fall in elevation between the leading and trailing parts of the head wave. Equating these expressions for p, we obtain

$$U_h = \sqrt{(2gh_1)}, \tag{6.1}$$

whence

$$U_h/\sqrt{(gh_1)} = \sqrt{2} = k_1. \tag{6.2}$$

In this expression, k_1 is a dimensionless velocity coefficient analogous to the Froude number and is also given by

$$k_1 = \left[\frac{(d-h_1)(2d-h_1)}{d(d+h_1)}\right]^{\frac{1}{2}}, \tag{6.3}$$

where d is the depth of the ambient medium. Comparing Eq. (6.2) for great depth and Eq. (6.3), we note that $k_1 \to 1/\sqrt{2}$ as $h_1/d \to 0{\cdot}5$.

No density term appears in Eq. (6.1) since the density difference between cavity and ambient medium equals the density of the medium. Putting ρ_1 as the density of the fluid now constituting the current, and ρ_2 as the density of the medium, it is easily found from revised expressions for the pressure that

$$U_h = k_1\sqrt{g\frac{\Delta\rho}{\rho_2}h_1}, \tag{6.4}$$

where $\Delta\rho$ is the difference of the densities (Fig 6.2).

Laboratory experiments made with solutions of salt or sugar, or with suspensions of plastic beads, reveal that the heads of gravity currents behave as predicted by Eq. (6.4), provided the Reynolds number for the flow is not small. Viscous drag at the solid boundary, however, causes the leading point of the head to be displaced a small distance away from the boundary (Fig 6.3). Relative to the ground, the motion inside the head is turbulent and strongly diverging, upwards towards the roof of the head and to a lesser extent downwards towards the solid boundary (Fig 6.3a). The largest velocities observed in the ambient fluid pushed aside by current are found above the broad, gently rising front of the head, a fact made use of by glider pilots flying at sea-breeze fronts. Fluid from the medium

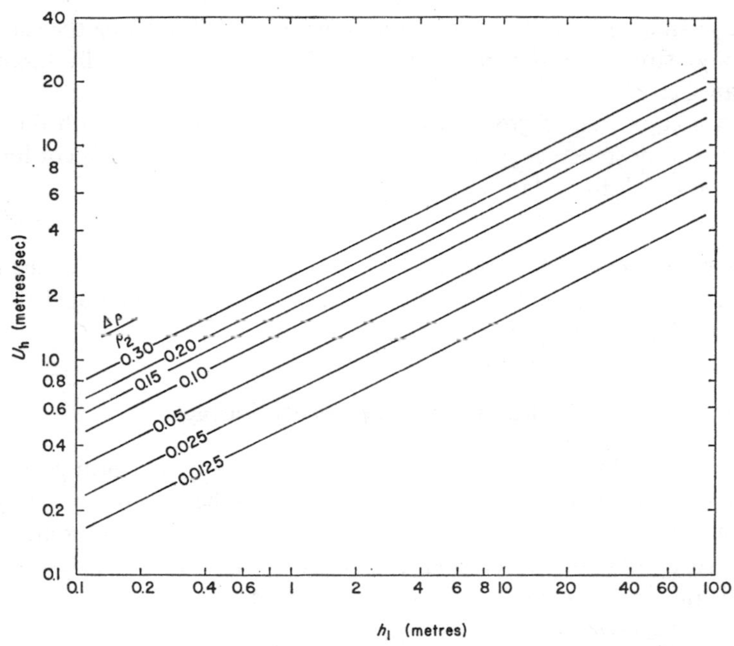

Fig 6.2 Velocity of the head of a gravity current as a function of thickness o flow and excess density ($k_1 = \sqrt{2}$).

becomes mixed into the current at the rear of the head where large, rapidly circulating vortices are seen. Directly in front of the head the medium is but little disturbed, though there may be some mixing into the head by way of vortices developed below the overhanging part of the interface. Streamlines drawn relative to the head reveal

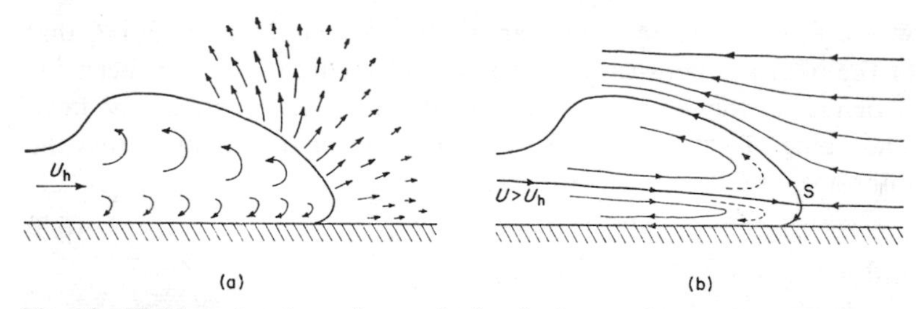

Fig 6.3 Fluid motion in and near the head of a gravity current. (a) Motion relative to the ground. The current vectors shown in the ambient medium are to be compared with the vector U_h (based on G. V. Middleton). (b) Motion relative to the moving head.

the leading point to be a stagnation point (Fig 6.3b). Along the stagnation streamline the local current velocity is substantially larger than U_h.

A comparison of gravity currents from the laboratory with those in the atmosphere suggests that the head becomes flatter according to the rough formula

$$l/h_2 = 0{\cdot}3\, Re^{\frac{1}{4}}, \tag{6.5}$$

where l is the length of the head, h_2 is the full thickness of the head, and Re is the Reynolds number based on h_2. It may be noted that $h_2 \approx 2h_1$.

6.3 Motion of a Steady, Uniform Gravity Current

By maintaining the supply of heavy fluid, a steady, uniform gravity current can be established behind the head. In the case of an underflow over a sloping bottom, the motion of the steady current is maintained by the body force, and not by the hydrostatic pressure-force as is the flow of the head. We therefore write

$$U_s = \sqrt{\left(\frac{8g\,\Delta\rho/\rho_2}{f}\right)}\sqrt{(h_3 S)}, \tag{6.6}$$

where U_s is the mean velocity of the steady current, f is the total resistance coefficient, h_3 is the thickness of the uniform flow, and S is the slope. Eq. (6.6) is analogous to the Chézy equation for open-channel flow, discussed in Ch. 4, but f is no longer the simple Darcy-Weisbach coefficient. We must also write

$$f = f_0 + f_i, \tag{6.7}$$

where f_0 expresses resistance at the bottom of the flow and f_i that at the interface between the current and medium. The coefficient f_0 depends on the Reynolds number and relative roughness of flow. The value of f_i depends chiefly on the extent of mixing across the interface, and appears to vary as

$$f_i \propto Re^{-3/5}, \tag{6.8}$$

and

$$f_i \propto \frac{U_s}{\sqrt{(gh_3\,\Delta\rho/\rho_2)}}, \tag{6.9}$$

where the term on the right in Eq. (6.9) is the densiometric Froude number. In laboratory experiments at moderate Reynolds numbers the ratio f_i/f_0 is close to unity. For large underflows on low slopes, however, the value of f_i is probably small compared to f_0 which is itself relatively small at the very large Reynolds numbers to be expected. Thus a large underflow will experience little fluid resistance, and the position of the velocity maximum in the velocity profile will lie in the upper half of the flow, close below the interface.

It is worth noting that on moderate and steep slopes, the velocity U_h is slightly smaller than the velocity U_s of the steady flow established behind the head. This implies the continual transfer of fluid from the uniform flow into the head, in order to make good losses from the head due to mixing in the lee.

6.4 Deposition from Experimental Turbidity Currents

The motion of a gravity current head is unsteady relative to the ground and rather like a surge. It is therefore hardly surprising that the motion is neither theoretically nor experimentally well understood. Our understanding is even less satisfactory when it comes to the transport and deposition of sediment by turbidity currents, though many valuable insights have come from laboratory experiments and limited observations in lakes and reservoirs.

The chief problem faced by the experimenter is the choice of model scale for the sediment carried in the current. Since the velocity scale of the experimental current must equal the square root of the length scale, in order to preserve Froude number similarity, it is necessary to scale down the sediment falling velocity proportional to the current velocity. Thus if we plan to reduce the velocity scale of the current to one-tenth, then the sediment falling velocity must be reduced by a like factor. It follows from the falling velocity laws that this can be done by changing either the particle size or the particle density or both. Geologists experimenting on turbidity currents have not always attended to the need for a correct scaling of the sediment used to give the current its excess of specific weight. Consequently many experimental results are of limited value.

A particularly important series of experiments has been made using for sediment a moderately well sorted mixture of plastic beads of density 1·52 g/cm^3 and mean diameter 0·18 mm. In each experiment, a suspension of the beads was suddenly released from a lock at one end of a long channel deeply filled with still water. The object of the experiments was to study the mode of deposition from the travelling bead suspension and the character of the vertical and lateral grading in the deposit.

The beds deposited in the channel varied in character with the initial concentration of beads in the lock. At low concentrations ($C \approx 0{\cdot}22$) the bed thickness decreased slightly with increasing distance from the lock, up to a distant point beyond which the decrease of thickness was rapid. Each bed was at every point graded vertically from coarse up to fine. The progressive change affected every percentile of the size distribution, and so was evident both from a measure of central tendency, such as the mean size, and from a measure of the coarsest grain size present. This type of grading is *distribution grading*. In the same beds there was a well developed lateral decrease of grain size with increasing distance from the lock. When the concentration of beads in the original suspension was high ($C \approx 0{\cdot}45$), the bed thickness remained constant over a large distance from the lock, though beyond a certain point there was again a rapid decrease of thickness. The vertical grading in these beds was of a different type, called *coarse-tail grading*, made evident only by an upward reduction in the size of the coarsest material present. The mean size of the beads remained constant as the distance from the base of the bed increased, and diminished with height only in the topmost part of the bed, where distribution grading occurred. There was very little consistent lateral size variation in beds showing coarse-tail grading. The sediment sorting improved upwards in these experimental graded beds, regardless of the type of grading shown.

Deposition from the flows varied in mechanism with the original concentration of beads, but in all cases began at, or a little upstream from, the rear of the head. In the low concentration flows the particles first deposited on the bed came to rest quickly after only a short tractional movement. The following beads were deposited in a particle-by-particle or layer-by-layer manner. Some of the later deposited particles experienced tractional movement over a substantial distance. A quite different sequence of events was observed for high concentration flows. The bed was laid down rapidly, in a bulk manner, and the interface between the bed and the flow overhead was scarcely detectable. As a consequence of the mode of deposition, the bed assumed a very loose packing and was sheared by the flow above into waves that severely disturbed particles to a depth of about two-thirds the thickness of the bed. Consolidation of the bed followed only on the disappearance of the wave motion.

Other experiments were made with dense suspensions ($C \approx 0{\cdot}3$) of quartz sand mixed with clay released at the head of either a long narrow trench or a short channel emptying below water into a deep broad basin. In the former case, each suspension was released suddenly from a lock at one end of the trench, as in the experiments using bead suspensions. The currents left behind in the trench beds that were graded vertically up from coarse to fine and horizontally from coarse near the lock to fine at a distance from it. The vertical grading ranged from coarse-tail grading in the lower part of a bed to distribution grading in the upper portion. The thickness of the beds tended to decrease exponentially with increasing distance from the lock. A similar decrease of thickness with increasing distance from the source was observed during experiments made in the broad basin, but as the supply of dense suspension was in these cases kept uniform for a relatively long period, the only grading observed was horizontal grading. Thus at each point along the profile the mean size and sorting of the bed was vertically constant, except at the very top, where grading occurred due to the final cessation of flow.

6.5 Initiation and Decay of Turbidity Currents in Recent Times

Although turbidity currents are frequently made and measured in the laboratory, and are sometimes observed and recorded from lakes and reservoirs, their occurrence on a grand scale and properties in the marine environment are matters for conjecture from circumstantial evidence.

Turbidity currents may well begin life as sediment slumps, though this plausible idea has been inadequately explored until recently, on account of ignorance of the requisite soil mechanics. The idea calls for slumps of a large size, to match in volume the thickness and extent of beds of coarse material known from the ocean floor. It also requires that the slump consist of a sufficiently weak material, and occur in a region of sufficiently large sustained slope, that acceleration can take place to a velocity great enough to permit mixing with the surrounding sea water and development of turbulence. Such a slump might ensue after an earthquake shock or period of depositional oversteepening.

The factors that theoretically favour slumps behaving as described above are a moderate to steep slope, a poorly consolidated or sensitive material (sand-grade deposits are not precluded), and an undrained failure of the material during which excess pore pressures can be sustained. Their influence can be seen by means of a simple analysis of forces in which the slump is treated as a rigid block on an infinite slope and viscous effects are neglected. We may write for the limiting equilibrium of the mass after an undrained failure has occurred

$$\frac{n}{\gamma'} = \frac{\cos\beta\tan\phi' - \sin\beta}{\tan\phi'}, \tag{6.10}$$

in which n is the gradient of excess pore pressure within the slump as measured downward perpendicular to the upper surface, γ' is the submerged bulk density of the slump, β is the slope angle, and ϕ' is

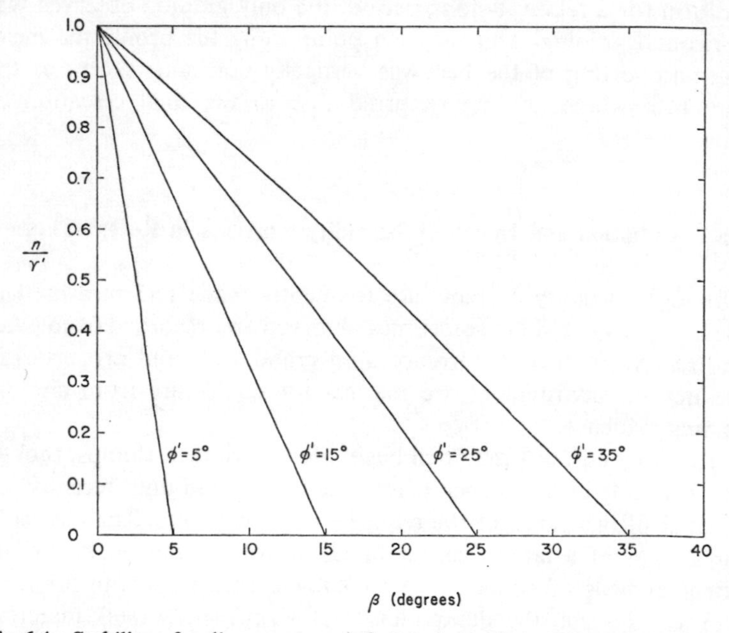

Fig 6.4 Stability of sediment on an infinitely long slope. The group n/γ' as a function of slope angle and angle of shearing resistance. The slope is unstable when n/γ' within the sediment mass lies to the right of the plotted curve for angle of shearing resistance.

the angle of shearing resistance under water of material in the slump. Eq. (6.10) is plotted in Fig 6.4 for different values of ϕ'. It will be evident that for a given β and ϕ', movement of the mass occurs only when n/γ' exceeds in value the expression on the right-hand side of Eq. (6.10). The velocity U of the slump becomes

$$U = \frac{gT}{\gamma}[\gamma' \sin \beta - (\gamma' \cos \beta - n) \tan \phi'], \tag{6.11}$$

in which γ is the unsubmerged bulk density of the slump material and T is the time elapsed after the start of motion. It follows that undrained slumps can attain large velocities rather quickly even on small slopes, as is required by Eqs. (6.4) and (6.6) if turbidity currents are to form from slumps (see Fig 6.2). For comparison, Fig 6.5 gives values of U calculated from Eq. (6.11) for $n = 0{\cdot}4$ g/cm^3, $\phi' = 10°$ and various values of β.

The marine environment at present favouring extensive slumping is the upper continental slope, with gradients typically between 1:5

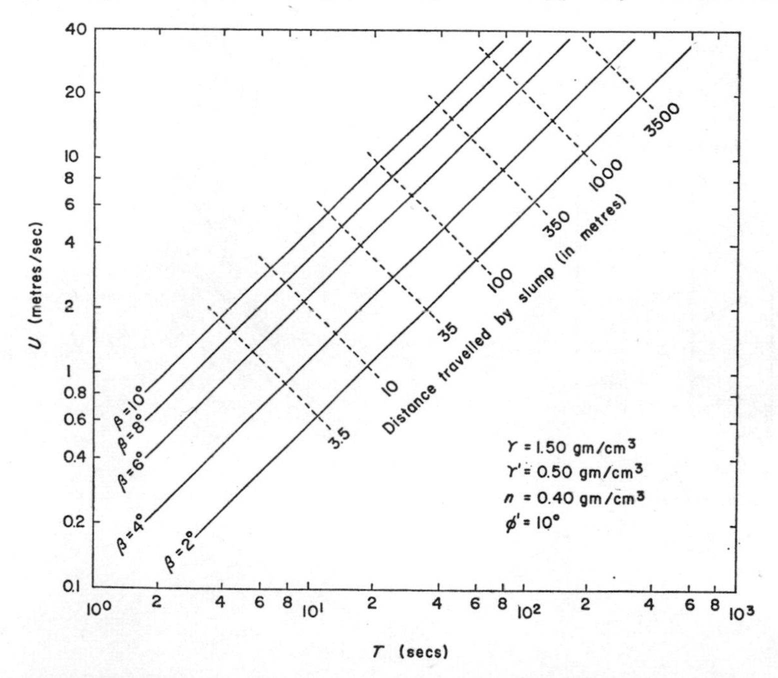

Fig 6.5 Velocity of advance of a frictionless slump, and distance travelled, as a function of time. The slumped mass is assumed initially at rest.

and 1:50. A representation of the slope and its neighbouring morphological regions appears in Fig 6.6. Below the slope is the continental rise, a broad area characterized by gradients in the range 1:100 to 1:700, which is not uncommonly dominated by partly or wholly coalescent submarine fans that slope away from the mouths of steep-walled canyons cut back into the continental slope. At the present time the continental slope is surmounted by the continental shelf. This region seldom attains depths greater than about 200 metres, but varies enormously in width, between a few kilometres and a few hundred kilometres. Generally speaking, however, the shelf is broad and gently sloping. Beyond the toe of the continental rise are smooth abyssal plains of gradient between 1:1000 and 1:10 000. Deep-sea channels can often be traced from the mouths of submarine canyons, across fans and on to abyssal plains. There are cases of channels that pass between ranges of abyssal hills to unite abyssal plains whose floors lie at different levels.

Historical submarine slumps have occurred at the heads of sub-

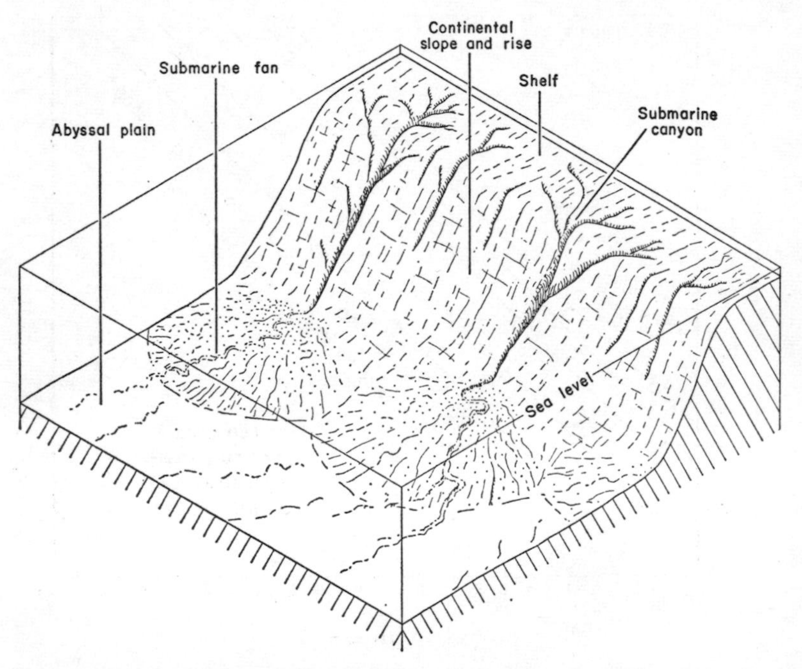

Fig 6.6 Submarine topography associated with turbidity current action. Vertical scale greatly exaggerated.

marine canyons, on the upper continental slope, and off the mouths of large rivers that empty on to relatively narrow shelves or near the shelf edge. All such sites as these are actually or potentially areas of relatively rapid deposition as well as relatively large slopes; many sites of these types occur in seismically active areas. Some of the recorded slumps can be attributed to depositional oversteepening, though others clearly followed earthquake shocks. The slumps occurred on slopes of gradient mainly between 1:6 and 1:30 and involved amounts of material ranging in volume between 3×10^5 and 7×10^{10} cubic metres. In several instances, notably that following the Grand Banks earthquake of 1929, the slump apparently became transformed into a turbidity current that swept far out over the rise and distant abyssal plains, to break in succession the submarine telegraph cables in its path and deposit a thick bed of coarse sediment. The evidence of cable-breaks is not without ambiguities, but where a freshly laid coarse bed is found in an area of broken and damaged cables, the action of a powerful current must be inferred. The timing of such breaks suggests that velocities in the range 5–10 metres per second are fairly characteristic of the currents, in accordance with the implications of Eq. (6.4).

It appears that today, typified by a high-standing sea level, submarine slumps occur at all frequently (annually or near-annually) only off the mouths of a few of the largest rivers. During the glacial periods of the Quaternary, however, when sea-level stood 100–150 metres below the present level, slumps were probably frequent on most parts of the continental slope and in most canyons. For rivers would then have emptied close to the modern shelf-edge and the areas of steepest slope. Thus rapid deposition of sediment, probably at the greatest rate closest to the land, would have occurred over great stretches of the continental slope. Intense slump activity chiefly due to depositional oversteepening can therefore be expected during glacial and other periods of lowered sea level. Since five main glaciations have occurred during the Quaternary period, we would expect evidence of extensive sediment deposition by turbidity currents against the present continental margins.

Submarine topography similar to that sketched in Fig 6.6 occurs along large stretches of the margins of the continents, and is particularly well developed off the mouths of big rivers, for example, the Rhône, Nile, Niger, Indus, Ganges, Magdalena and Mississippi. The morphological assemblage—a gently sloping channelled fan between a smooth flat plain and a steep dissected slope—is remarkably like the well-known assemblage of basin, alluvial fan and mountain front known from the land. The similarity extends to many details of form. Thus, the channels which cross submarine fans display, like their counterparts on alluvial fans, meanders and bordering levées. The submarine fans are underlain by beds of sand, a few centimetres to metres in thickness, in alternation with layers of hemipelagic mud. Many of the sands are rich in terrigenous material and often contain shallow-marine faunas and even land macro-floras, though found in water depths of several thousand metres. Other sand beds, equally out of place in the deep waters of the fans and basins, consist chiefly of carbonate-mineral particles of biogenic and chemical origins (including blue-green algae) originating in sunlit shallows. Even laminations and cross-lamination, indicative of currents ranging widely in power, are generally present in these sand beds, which commonly are graded vertically from coarse up to fine. Graded sand layers are numerous amongst the deposits underlying abyssal plains though they are generally thinner, finer grained, and less frequent than the beds proved in the fans. Acoustic profiling and careful correlation of bottom cores has shown that the sand beds of abyssal plains and submarine fans commonly extend individually over areas measuring hundreds and even thousands of square kilometres. Thus the seascape, and the associated sediments, point fairly conclusively to the operation of submarine 'rivers' capable of swiftly carrying large quantities of coarse sediment from shallow to deep water. These rivers, like their counterparts on land, appear to have deposited most of their load at the foot of the steepest slope but to have carried some to deep far-flung basins.

The existence of submarine fans implies that, after an initial acceleration, the surge-like current which descends the slope proceeds to lose power once it has spilled on to the fan. Since the fan declines progressively in slope away towards the abyssal plains, the effect on the current of a progressive reduction of the down-slope body force is added to the effect of reduction of excess specific weight due to mixing with sea water. Sediment deposition necessarily ensues, itself contributing to the steady reduction of the driving forces.

Because sediment transport in suspension depends on turbulence, a model of turbidity current deposition can be made by considering the effect on a discrete mass of turbulent fluid carrying sediment of a decay of turbulence intensity. The turbulence may for simplicity be assumed to be isotropic. It may also be assumed, in the context of turbidity currents generated from slumps, that sediment particles of a broad range of sizes are uniformly available. By the assumptions made and the turbulence decay laws, it can be shown that the rate of deposition from the mass of particles of falling velocity V_0 is

$$E = k_2 mhNV_0 T^{(m-1)} \exp(-k_2 V_0 hT^m), \qquad (6.12)$$

in which E is the rate of deposition (number of particles per unit area per unit time), k_2 is some constant, m is an exponent in the decay laws ($1 < m < 2{\cdot}5$), N is the number of particles of falling velocity V_0 in unit volume of the mass at the start of decay, h is the thickness (small) of the turbulent layer, and T is time after the start of decay; the notation $\exp(x)$ means e^x, e being the base of natural logarithms. Since increasing time of decay means increasing distance of transport, the equation can be developed to show that the bed deposited by a turbidity current decreases in thickness with increasing distance from the point, presumably the fan apex or toe of the continental slope, where power first is lost.

By an extension of Eq. (6.12), the relationship between the vertical height y, measured above the base of a bed deposited from a turbidity current, and the grain size at that height becomes

$$y = k_3 (D^5_{max} - D^5_{mod}), \qquad (6.13)$$

in which D_{max} is the diameter of the largest particles present in the original size distribution ($T = 0, y = 0$), D_{mod} is the modal diameter of the particles deposited at height y, and k_3 is a constant. This equation states that D_{mod} declines with ascending height above the base, whence the bed is graded. If the original size distribution of sediment in the mass of turbulent fluid is now supposed non-uniform, Eq. (6.13) must be revised to read

$$y = k_4 (D^{(2n+5)}_{max} - D^{(2n+5)}_{mod}), \qquad (6.14)$$

in which k_4 is a new constant and n is a constant describing the size distribution. If $n = -2$, we find that y and D_{mod} are linearly related. The curve of y on D_{mod} is, however, convex upward for $n > -2$ (but $\neq -1$), though concave upward for $-2{\cdot}5 < n < -2$. If, in Eq. (6.14), we set $D_{mod} = 0$, then providing $n > -2{\cdot}5$ we obtain

$$Y = k_4 D^{(2n+5)}_{max}, \qquad (6.15)$$

in which Y is the total thickness of the graded bed. Interpreting

D_{max} as the largest particle diameter in the bed, assumed to occur at the base, the equation says that bed thickness and coarseness are directly proportional in turbidity current deposits.

The analysis behind Eqs. (6.12–15) leads to the conclusion that a turbidity current, in its decaying phase, deposits a bed that thins along the current path, shows an upward grading at any point from coarse to fine, and decreases in thickness as the maximum particle diameter preserved grows smaller. The analysis may chiefly be criticized on the grounds that it neglects the generation of turbulence in turbidity currents, and thus assigns to the decay time and length scales that are too short. It is also entirely kinematic and ignores the possibility that a turbidity current can be virtually self-sustaining.

We need not suppose that a turbidity current is incapable of eroding the substrate once, with loss of power, it has begun to deposit sediment. For it arose from the bead experiments that deposition began at, or a little behind, the rear of the head. The head itself is therefore a region potentially of erosion of the substrate, even though deposition on the substrate occurs to the rear. The extent of erosion at a point in the current path will depend on the shear stress exerted by the current, the erosion resistance of the substrate, and the transit time of the head relative to the point. Since the substrate is ordinarily mud, any substantial erosion should normally be expressed as flute marks.

In Eq. (2.26) we have an expression relating flute amplitude to duration of current action and rate of erosion on the bed surrounding the mark. It is of interest to develop this equation further, since we can show that flute amplitude is a key to the scale and velocity of the parent turbidity current. We may assume that the boundary layer in the head of the turbidity current is turbulent and grows like the boundary layer on a flat plate, and that from Ch. 1 the general erosion law is

$$E = k_5 \tau_0^n, \qquad (6.16)$$

in which E is the rate of erosion, k_5 is a constant depending on the erosion resistance of the substrate, τ_0 is the boundary shear stress and n is a positive exponent. From Eq. (2.26), the erosion law, and the turbulent boundary layer equation for shear stress, we may write for flute amplitude

$$A = \frac{1}{\gamma} k_5 (k_1 - 1) T \left[\frac{0{\cdot}228 \rho_1 U^2}{(\log_{10} Re)^{2{\cdot}58}} \right]^n, \qquad (6.17)$$

in which γ is the bulk density of the substrate, k_1 is the erosion rate ratio of Eq. (2.26), U lies in the range $U_h < U < 2U_h$, ρ_1 is the density of the current and *Re* is the Reynolds number based on U_h and the head length. The transit time T of the current is obtained from l and U_h as given by Eqs. (6.4) and (6.5). Eq. (6.17) states that, for a given substrate, the amplitude of a flute increases with ascending thickness and excess density of the current. Furthermore, when experimental values of k_1, k_5 and n are used in the equation, amplitudes consistent with experience are obtained. To conclusions afforded by Eqs. (6.12–15), we may now add the expectation that the flutes formed by a turbidity current decrease in size with increasing distance along the current path.

It should be stressed that whereas turbidity currents are probably the most important means whereby coarse terrigenous and shallow-water carbonate sediments reach ocean depths, there exist at these same depths persistent currents locally powerful enough at the bed to rework introduced coarse material. These currents, discussed in Ch. 5, manifest normal oceanic circulations. The evidence for their role is the occurrence on the ocean bed of active ripples and dunes where sands abound and, where there is mud, of features denoting scour or spatially variable deposition.

6.6 Ancient Turbidites

The term turbidite begs the question, of course, when applied to bodies of rock, but it is used by geologists in a consistent manner for a distinct sedimentary facies, with the following general properties. Turbidite formations are ordinarily a few thousand metres thick and consist of large numbers of coarse and fine grained beds vertically alternating on a small to medium scale. The coarse beds are generally of sand grade material and often vertically graded from coarse up to fine. Their soles are invariably sharp and often decorated with flute or tool marks, load casts, or organic markings. The fine beds pass up gradually from the coarse ones and vary from silt to clay grade. The coarse beds are remarkable for their extreme lateral persistency without substantial change of thickness or internal characters. Faunally, the turbidite facies is distinguished by deep water and offshore pelagic forms in the fine beds, and displaced terrestrial to shallow-marine elements in the coarse ones. A fauna of any sort is, however, rare in all but bioclastic turbidites. The organic markings sometimes found underneath the coarse beds are chiefly animal tracks and trails produced at the sediment-water interface prior to arrival of the coarse bed. They indicate that conditions were normally calm, so that benthonic animals had, by and large, no need to shelter in the substrate. Animals which live by burrowing through and eating the bottom sediment are, however, fairly common in the deep sea.

Good examples of turbidite formations are described from geosynclinal successions as exemplified by the Palaeozoic sequences of Wales and Southern Scotland; the Mesozoic and Neogene Flysch of the French and Swiss Alps, the Apennines, and the Carpathians; the Ordovician and Devonian of the Appalachian region of the USA; and the Upper Palaeozoic succession of the Ouachita Mountains, Oklahoma and Arkansas.

The turbidity current, conceived from experimental studies and investigations of the sea bed as a surge, displays an early phase of acceleration and a later stage of deceleration and decay. Deposition from the current may not be restricted to the later phase, but is certainly characteristic of it. On account of the unsteady depositional process, we expect in ancient turbidite formations to find correlations between lithological attributes in a given vertical profile (i.e. at a fixed point from source), and also within the beds systematic lithological variations along the current path (i.e. with increasing distance from source). The path itself, as revealed by current directional structures, should be relatively homogeneous.

The chief current-directional structures of turbidites are flute and tool marks, preserved as moulds on the soles of the coarse beds, and cross-lamination where coarse and fine parts grade. Mapping of sole markings on a local or sub-regional scale has shown that the majority of turbidite formations were derived from a single source direction. In most cases the currents varied in direction by no more than 40°–50° during the time necessary to deposit the formation, and in certain cases the range is only half this amount. It is rare for a directional range greater than 150° to be reported. Bimodal current distributions with modes about 90° apart are sometimes found, though distributions with directly opposed modes are practically never recorded. The cross-lamination in turbidites is seldom mapped, as it is an intrinsically variable structure. Some of it may record ordinary oceanic circulations rather than turbidity currents.

The vertical and lateral variation to be observed in typical turbidite units, each composed of a coarse grained bed overlain by a fine one, is sketched in Fig 6.7. The completely developed turbidite unit consists of five divisions labelled upwards from *A* to *E*. The divisions *A* to *C* represent the coarse part of the unit, while divisions *D* and *E* represent the fine grained bed. It will also be noticed that divisions *A* to *E* appear in sequence horizontally along the current path at the bottom of the bed. The length scale associated with the horizontal trend is much less certain than the rate of vertical variation, but the change occurs over distances probably between a few tens and a few hundreds of kilometres according to sedimentary basin.

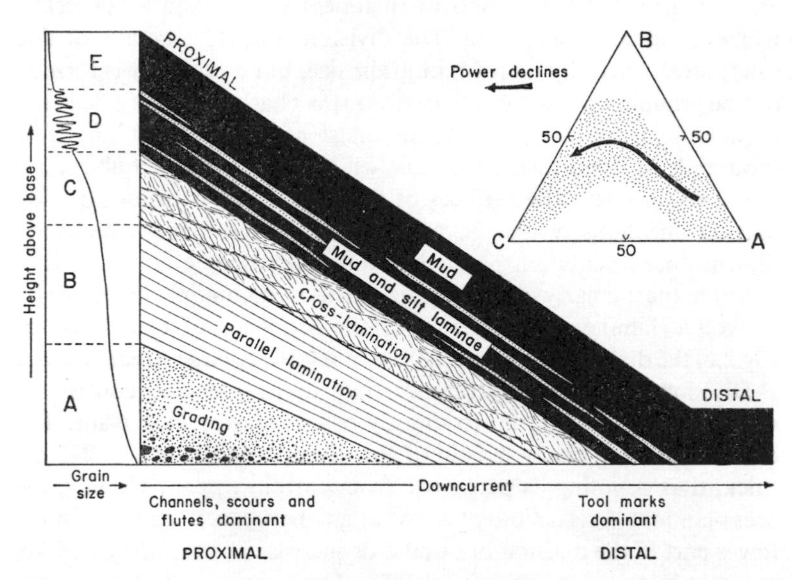

Fig 6.7 Vertical and lateral (downcurrent) variation in an ideal turbidite bed (partly after R. G. Walker).

The graded division *A* is the coarsest part of the turbidite unit. In the case of a terrigenous turbidite, it consists typically of medium to coarse sand graded from coarsest at the bottom to finest at the top. At the bottom there is commonly a thin layer in which very coarse sand grains, granules and even small pebbles are concentrated. Large chunks of mud, sometimes distorted or even partly rolled up, may also be found in the division. Log-log-plots of grain size for this coarse basal element against total coarse bed thickness yield straight lines of slope in the range 0·36–0·92. Distribution grading is much more frequent in division *A* than coarse-tail grading. Rectilinear, concave upward, and convex upward grading curves (arithmetic scales) have all been reported from division *A* of turbidites. The division typically is massive and without internal layering, though occasionally impersistent stringers of particles coarser or finer than average are found. It is commonly several decimetres thick. Division *A* has been interpreted to mean deposition from a flow of very large stream power; shearing within the bed may have been a factor in determining its massive character.

Division *B*, of parallel laminations, is ordinarily thinner and finer grained than division *A*, and in thickness varies from a few centimetres to as much as 50 cm. The division consists generally of fine sand, often vertically uniform in grain size, but commonly is formed of fine grading up into very fine sand. It is characterized by delicate, parallel laminations whose lateral persistence is several hundreds or thousands of times their thickness, which ranges between about 0·5 and 2 millimetres. The surfaces of the laminae on occasions display parting lineations, while the grains in the layers have a preferred dimensional orientation parallel to any macroscopic lineation. The division, particularly in its upper part, is commonly affected by convolute lamination. In terms of the sequence of bed forms of Fig 2.6, the division represents a flow of relatively large stream power.

Division *C* is normally thinner than division *B* and consists of very fine sand grading up to muddy coarse to very coarse silt. The cross-lamination characteristic of the division occurs in sets 2–3 cm thick, thus denoting its parentage from current ripples. Usually, the cross-laminated sets climb at a low angle, between 5° and 15°, in the lower part of the division but upwards increase in steepness of climb, commonly to an angle of 45°–55°. This upward change, in the context of a turbidity current of gradually waning power, can be explained in the light of Eq. (2.9) to mean a rate of deposition of sediment from the current decreasing less rapidly with time than the rate of bed load transport. Since grain size also decreases upward within division *C*, often to coarse or very coarse silt at the top, it may well be that bed load movement eventually almost ceased, in anticipation of the predominant settling from suspension indicated by division *D*. In parallel with the increasing steepness of climb, the relationship of one cross-laminated set to another in division *C* changes generally from erosional in the lower part of the division to gradational in the upper part, where laminae are preserved on ripple stoss-sides. The ripples, too, commonly change in form upwards. They are frequently long and low in the lower part of the division but short and tall in the upper portion. At the top the ripples are often seen as symmetrical undulations. Bedding planes exposed in division *C* show ripples varying in plan from long and straight-crested to linguoid. In occasional turbidite units, cross-bedded sandstones in units a decimetre or two thick occur above the parallel laminations of division *B* and below cross-laminated deposits. These more thickly cross-stratified sediments may record dunes, which are formed at somewhat larger stream powers than ripples (Fig 2.6). Division *C*, like division *B*, is commonly affected by convolute lamination. In certain cases it appears that the convolutions grew at the sediment-water interface as deposition proceeded. The division clearly records rapid accumulation from a flow of small to moderate stream power (Fig 2.6).

Division *D* is clean silt, muddy silt, or silty clay in thin parallel laminae. Occasional lenses of clean silt, internally cross-laminated, are found in the division. The succeeding division *E* consists of clayey silts and silty clays that are massive and blocky in the lower part but fissile in the upper layers. Much if not all of division *E* is hemipelagic in origin.

In addition to the above internal changes, there are systematic lateral variations to be observed in the character of, and structures recorded from, the soles of the coarse beds (Fig 6.7). Where the coarse beds are thick, the soles frequently show irregular scours, washouts or small channels. Flute moulds are generally common and close-packed; they often reach a large size, up to 15 cm in amplitude and 1 metre in length. The tool marks present are also generally large, but scarcely any surface tracks and trails are recorded. Quite commonly, a group of such turbidite units occupies a steep-walled channel whose depth may reach several tens of metres. As the coarse beds of turbidites become thinner, however, scours and channels become smaller and less frequent, eventually disappearing, and flute and tool marks are found to be less large. Organic sole markings become fairly common, though not all represent surface forms. Where the coarse beds are thinnest, tool and organic markings predominate over markings due to scour. Washouts are now small and infrequent, but tracks and trails abound. Very often, the soles of the coarse beds prove to be perfectly plane, lacking even the most delicate of tool marks. These soles overlie surfaces that may not have experienced erosion by the advancing turbidity currents.

The changes discernible in turbidite units collectively are more easily dealt with. In the direction of bed thinning, there is a gradual increase in the ratio of the thickness of fine grained rocks (divisions *D* and *E*) to the thickness of coarse beds (Fig 6.7). Where the coarse beds are thickest, the argillaceous deposits are relatively very thin and may even be absent, the coarse graded sequences being amalgamated together. A commonly observed vertical pattern in turbidite formations is a 'bundling' together of units of about the same thickness, so that relatively thick units, a few tens together, alternate in groups with relatively thin ones. The general character of the turbidite units making up a local section can be described by determining the percentage of beds in the section which begin with division *A*, or division *B*, or division *C* (Fig 6.7). These percentages are easily graphed on a triangular diagram. As is rather to be expected, the plotted points move progressively from the *A* vertex to the *C* vertex as the formation is explored at increasing distances along the current path. The percentages denote the proximality of the section relative to the source of the turbidite units.

It will be clear from the above that there is a very fair consistency between the inferences that can be drawn from ancient turbidities and our knowledge of turbidity currents as founded on theory, experiment and the investigation of the modern sea floor. We appear to be dealing with currents able to carry coarse sediments in large quantities from shallow to deep water. Most seem to begin as sediment slumps in areas of relatively steep slope. Deceleration of the current begins once it reaches gently sloping parts of the sea bed. In the course of the ensuing deposition, the current gradually loses power with increasing distance of travel along its path and with increasing time at a point on the path. The result is a graded unit which becomes finer grained upwards and thinner and finer grained in the direction of the current. However, as our inferences regarding turbidity currents grow in scope and depth, so the need for direct observation of the natural currents becomes more acute.

READINGS FOR CHAPTER 6

There exists no standard work in which gravity and turbidity currents are satisfactorily treated from a geological standpoint, though the matter is important to atmospheric as well as oceanic processes. The following papers, however, provide an introduction to the theory and physical behaviour of gravity and turbidity currents:

BAGNOLD, R. A. 1962. 'Autosuspension of transported sediment: turbidity currents.' *Proc. R. Soc.*, **A, 265**, 315–319.
BELL, H. S. 1942. 'Density currents as agents for transporting sediments.' *J. Geol.*, **50**, 512–547.
BENJAMIN, T. B. 1968. 'Gravity currents and related phenomena.' *J. Fluid Mech.*, **31**, 209–248.
DALY, B. J. and PRACHT, W. E. 1968. 'Numerical study of density-current surges.' *Physics Fluids*, **11**, 15–30.
KUENEN, P. H. and MENARD, H. W. 1952. 'Turbidity currents, graded and non-graded deposits.' *J. sedim. Petrol.*, **22**, 83–96.
KUENEN, P. H. and MIGLIORINI, C. I. 1950. 'Turbidity currents as a cause of graded bedding.' *J. Geol.*, **58**, 91–127.
MIDDLETON, G. V. 1966(7). 'Experiments on density and turbidity currents.' Parts I–III. *Can. J. Earth Sci.*, **3**, 523–546, 627–637, **4**, 475–505.

The papers of Kuenen are of particular historical importance, even though as regards their experimental aspects they have to a large extent been superseded by the more recent work of Middleton. Benjamin gives the simplest theoretical model of a gravity current.

Many papers have now been written on turbidity currents as the cause of graded sand layers in the bottom deposits of the present seas and oceans. The following are instructive as well as representative:

CONOLLY, J. R. and EWING, M. 1967. 'Sedimentation in Puerto Rico Trench.' *J. sedim. Petrol.*, **37**, 44–59.
EMERY, K. O. 1960. *The Sea off Southern California. A modern Habitat of Petroleum.* John Wiley & Sons, Inc., New York, 366 p.
HEEZEN, B. C. 1963. 'Turbidity currents.' In (M. N. Hill, ed.) *The Sea.* Interscience Publishers, New York, vol. 3, p. 742–775.
HOLTEDAHL, H. 1965. 'Recent turbidites in the Hardangerfjord, Norway.' In (W. F. Whittard and R. Bradshaw, eds.), *Submarine Geology and Geophysics.* Butterworths Scientific Publications, London, p. 107–140.
HOUBOLT, J. J. H. C. and JONKER, J. B. M. 1968. 'Recent sediments in the eastern part of the Lake of Geneva (Lac Leman).' *Geologie Minjb.*, **47**, 131–148.
RYAN, W. B. F., WORKUM, F. and HERSEY, J. B. 1965. 'Sediments on the Tyrrhenian Abyssal Plain.' *Bull. geol. Soc. Am.*, **76**, 1261–1282.
SHEPARD, F. P. 1961. 'Deep-sea sands.' *Rep. 21st int. geol. Congr.*, Pt. 23, 26–42.
SHEPARD, F. P. and EINSELE, G. 1962. 'Sedimentation in San Diego Trough and contributing submarine canyons.' *Sedimentology*, **1**, 81–133.

The connection between turbidity currents and sediment slumping is discussed in a valuable article by:

MORGANSTERN, N. R. 1967. 'Submarine slides, slumps, and slope stability.' In (A. F. Richards, ed.), *Marine Geotechnique*, University of Chicago Press, Chicago, 327 p.

The following articles, some theoretical in basis, cover most of the important lithological features of rocks attributed to the turbidite facies:

ALLEN, J. R. L. 1970. 'The sequence of sedimentary structures in turbidites, with special reference to dunes.' (to be published in *Scottish Journal of Geology*.)

DZULYNSKI, S. and WALTON, E. K. 1965. *Sedimentary Features of Flysch and Graywackes*. Elsevier Publishing Co., Amsterdam, 274 p.

POTTER, P. E. and SCHEIDEGGER, A. E. 1966. 'Bed thickness and grain size: graded beds.' *Sedimentology*, **7**, 233–240.

SCHEIDEGGER, A. E. and POTTER, P. E. 1965. 'Textural studies of graded bedding: observation and theory.' *Sedimentology*, **5**, 289–304.

WALKER, R. G. 1965. 'Origin and significance of the internal sedimentary structures of turbidites.' *Proc. Yorks. geol. polytech. Soc.*, **35**, 1–29.

WALKER, R. G. 1967. 'Turbidite sedimentary structures and their relationship to proximal and distal depositional environments.' *J. sedim. Petrol.*, **37**, 25–43.

A vast number of studies have been made in the past two decades on formations representative of the turbidite facies. The following papers are representative both of the geological range of the facies and the differences of approach between authors:

ALLEN, J. R. L. 1960. 'The Mam Tor Sandstones: a "turbidite" facies of the Namurian deltas of Derbyshire.' *J. sedim. Petrol.*, **30**, 193–208.

ANGELUCCI, A. and others, 1967. 'Sedimentological characteristics of some Italian turbidites.' *Geologica Romana*, **6**, 345–420.

CUMMINS, W. A. 1957. 'The Denbigh Grits; Wenlock graywackes in Wales.' *Geol. Mag.*, **94**, 433–451.

DZULYNSKI, S. and SLACZKA, A. 1958. 'Directional structures and sedimentation of the Krosno Beds (Carpathian flysch).' *Annls. Soc. géol. Pologne.*, **28**, 205–259.

ENOS, P. 1969. 'Anatomy of a flysch.' *J. sedim. Petrol.*, **39**, 680–723.

MCBRIDE, E. F. 1962. 'Flysch and associated beds of the Martinsburg Formation (Ordovician), Central Appalachians.' *J. sedim. Petrol.*, **32**, 39–91.

MCBRIDE, E. F. 1966. 'Sedimentary petrology and history of the Haymond Formation (Pennsylvanian), Marathon Basin, Texas.' *Rep. Invest. Bur. econ. Geol. Univ. Tex.*, No. 57, 101 p.

TEN HAAF, E. 1959. *Graded Beds of the Northern Apennines*. Thesis University of Groningen, 102 p.

WARREN, P. T. 1963. 'The petrography, sedimentation and provenance of the Wenlock rocks near Hawick, Roxburghshire.' *Trans. Edinb. geol. Soc.*, **19**, 225–255.

WOOD, A. and SMITH, A. J. 1959. 'The sedimentation and sedimentary history of the Aberystwyth Grits (Upper Llandoverian).' *Q. Jl. geol. Soc. Lond.*, **114**, 163–195.

In the books of Dzulynski and Walton (above) and Potter and Pettijohn (see Readings for Chapter 2), a number of regional studies of turbidites are examined.

CHAPTER 7

Glaciers and Glacial Deposits

7.1 General

Ice by nature is confined to cold climates where it is found in masses of several different forms and a huge range of sizes. From the standpoint of geological effectiveness, we may distinguish glacier ice, capable of transporting debris over or off the land, from sea ice, which seldom if ever has this capability. A glacier is a body chiefly of ice consisting of recrystallized and compacted snow, with some refrozen melt water and recrystallized hoar frost, that lies wholly or mainly on the land. Sea ice, generally found in the form of a floating ice shelf, is made up partly of ice consisting of frozen sea water and partly of ice formed from compacted snow. A body of sea ice generally has no connection with the land, though it may ground locally on the sea bed.

In high latitudes ice occurs on the land from sea level upwards, and is also found on the sea surface. But as equatorial regions are approached, areas of cold climate are restricted to increasingly higher altitudes, and so the lower limit of permanent ice, which may be taken as approximately the same as the climatic snowline, rises to higher and higher elevations. The climatic snowline does not rise uniformly with decreasing latitude, however, as its elevation is controlled by both mean annual temperature and amount of precipitation. For a given mean temperature, the snowline is increasingly depressed with ascending amount of precipitation. In equatorial regions the snowline is found between 4000 and 5500 metres above sea level but at 45° latitude lies at 1800–4000 metres. The climatic snowline at 60° from the equator is as low as 1100 metres above sea level.

At present approximately $1{\cdot}5\times10^7$ km^2 of the surface of the earth are covered by ice, the largest individual masses being found, as expected, in polar regions (Fig 7.1). By far the largest ice-covered region is Antarctica, with an area of about $1{\cdot}27\times10^7$ km^2. In the Northern Hemisphere, Greenland, with an ice-covered area of approximately $1{\cdot}8\times10^6$ km^2, comes a poor second. Most other ice-covered areas are very small compared even with the Greenland ice-sheet. The total volume of present ice is about $3{\cdot}3\times10^7$ km^3.

It is estimated that during the Pleistocene period the area covered by ice was as much as 3·2 times larger than the present area. The

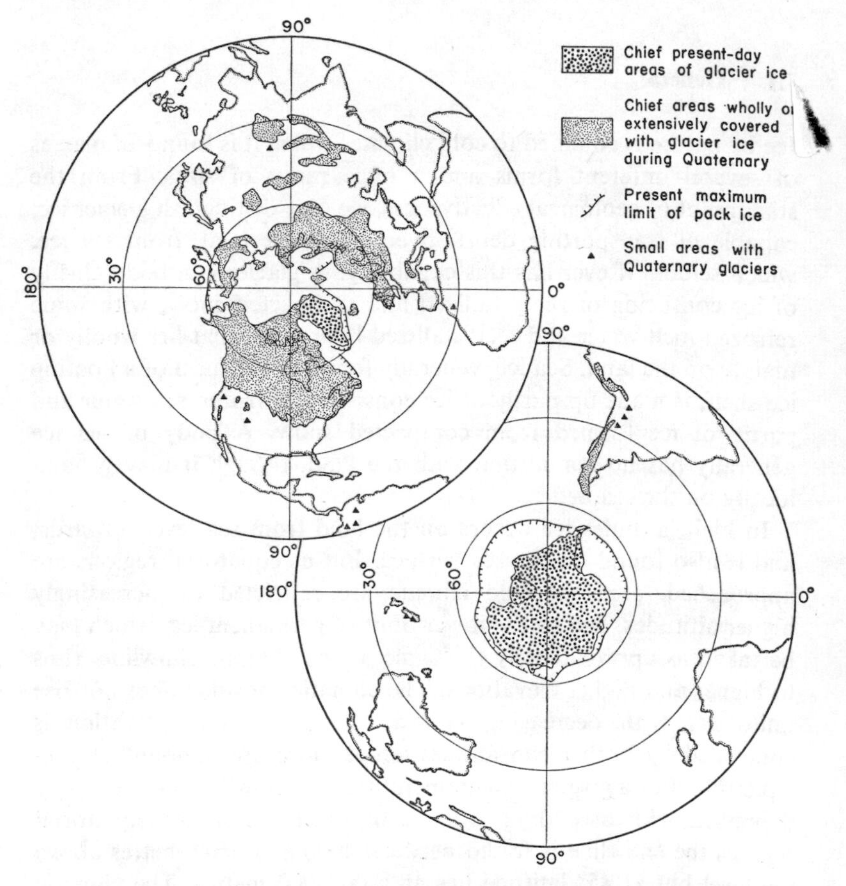

Fig 7.1 Generalized world distribution of glacier ice at the present day and during the Quaternary. The very numerous smaller areas of glacier ice are not shown.

volume of Pleistocene ice may have amounted to $1{\cdot}0 \times 10^8$ km^3. Five episodes of glaciation can be recognized in this period, and during each episode a considerable part of North America, north-west Europe and Asia was overwhelmed (Fig 7.1). These ice-sheets and glaciers have left behind characteristic landscape forms and series of deposits that are sufficiently youthful and fresh as to be amenable of rather complete documentation and interpretation. There is, moreover, good if fragmentary evidence of glaciation more than once before Cambrian times and again in the Permo-Carboniferous period. Claims have been made for glaciation at other pre-Pleistocene dates. The sediments laid down directly by the action of ice are known as till or, if cemented, as tillite. Closely associated with these in time and space are a variety of ice-contact stratified drifts. The tills and drifts of Pleistocene age are of considerable economic importance, in addition to being of intrinsic interest, if only because they give rise to some major soil groups in present-day temperate regions.

7.2 Some Physical Properties of Ice

When pure water at ordinary atmospheric pressure is cooled to 0°C it changes into the solid-state form known as ice. However, the freezing point of water diminishes by approximately 1°C for each additional 140 atmospheres pressure. At −70°C ice takes on a cubic structure and under very high pressures assumes a variety of poorly known crystalline structures. In the temperature and pressure ranges of practical interest, however, ice crystallizes on the hexagonal system, as may be inferred from the macroscopic forms of snow flakes.

In hexagonal ice, the only crystallographic species with which we shall be concerned, each oxygen atom is surrounded by four approximately equally spaced oxygen atoms arranged at the vertices of a tetrahedron. The basal plane of this lattice contains three oxygen atoms arranged to form an equilateral triangle. Perpendicular to the basal plane is a bond between two oxygen atoms that is parallel to the *c*-axis of the crystal, one of the oxygen atoms lying at the centre of the tetrahedron. The hydrogen atoms lie on the bonds between the oxygen atoms, though it appears that each hydrogen atom is closer to one of the oxygen atoms it is bonded with than the other.

The ice lattice is very open, for the spacings between the atoms are rather large compared with the sizes of the atoms themselves. The density of ice, 0·9168 g/cm^3 at 0°C, is even smaller than that of water. Another consequence of the structure of the lattice is that a single hexagonal crystal of ice is anisotropic as regards many physical properties. The crystal is isotropic in all directions perpendicular to the *c*-axis but is anisotropic as regards properties measured parallel and perpendicular to the *c*-axis. Thus a single hexagonal ice crystal has no less than five distinct elastic parameters, though it is also capable of plastic flow on account of the existence of glide planes parallel to the basal plane of the lattice. A shear strain and a resulting slip occur quite readily if a shear stress is applied to a crystal parallel to this basal plane. Of course, anisotropy would be apparent in polycrystalline ice only if the constituent crystals showed a preferred crystallographic orientation. However, most natural ice displays some degree of ordering of the crystals, greater in sea or lake ice than in glacier ice, as the *c*-axes tend to occur normal to the surface of accumulation. Natural ice in bulk is therefore anisotropic as regards many physical properties.

7.3 Glacier Types

Useful classifications of glaciers can be made on the basis of their thermal characteristics and geomorphological features.

The thermal properties of glaciers to a considerable extent measure glacier activity in terms of speed of movement and intensity of bed erosion. Two main thermal categories of glacier are recognized, temperate (warm) and polar (cold), though an individual glacier can be cold in one part and warm in another. A temperate glacier is a relatively warm one of which the ice is throughout at or very close to the pressure melting-point, except in winter when the upper layers become relatively cold. In a temperate glacier water can accumulate freely, in surface lakes and subsurface crevasses and tunnels. Meltwater occurs at the sole of the glacier, so facilitating basal slip. The polar glacier is, on the other hand, cold, and the temperature of its ice lies well below the pressure melting-point. Polar glaciers lack englacial or subglacial drainage, though during the brief summer melting season they commonly support large surface lakes and streams. The ice of a polar glacier very probably is frozen to the bed, so that there is no slippage between the basal ice and the bed material; moss-covered boulders have been found at the base of one cold glacier. Thus polar glaciers flow less freely than temperate ones because they lack the basal lubricant of meltwater and are thus firmly attached to their beds. It is also probable that polar glaciers erode at their beds far less readily than temperate ones.

Geomorphologically, the following classification covers the more important known types: (i) cirque glaciers, (ii) valley glaciers of Alpine type, (iii) valley (outlet) glaciers, (iv) piedmont glaciers, (v) ice-shelves and floating ice-tongues, (vi) mountain ice-caps, (vii) lowland ice-caps, (viii) ice-sheets. These different forms are shown schematically in Fig 7.2.

The cirque glacier, a type common in the Alps and Scandinavia, is a relatively small mass of ice that lies in a hollow or rock basin deeply sculptured back into the flank of a mountain. Smaller than cirque glaciers, but representing an early stage in their development, are niche and cliff glaciers which develop in grooves in the rock or

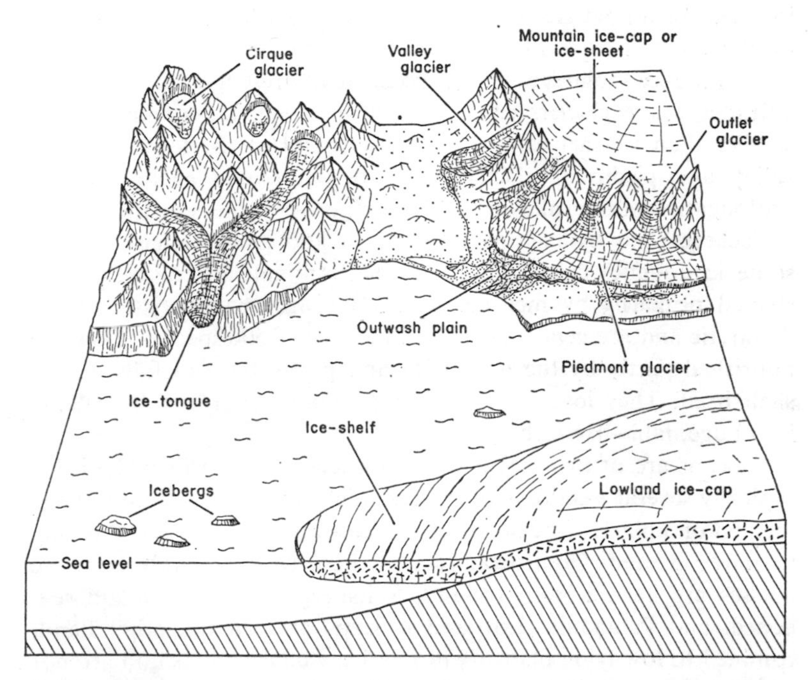

Fig 7.2 Schematic representation of the main geomorphological types of glacier.

at the feet of steep cliffs. Cirque glaciers are a few tens to a hundred or so metres thick.

Valley glaciers are ice streams confined within mountain walls. They are especially well known as they are relatively small in size and easy of access; an ice thickness of a few hundred metres is fairly typical. The Alpine type, which of course is not confined to the Alps, is fed at the head from a number of coalescent cirque glaciers but has few or no other tributaries. Some valley glaciers, however, have numerous large tributaries, as in Spitsbergen. Another variety of valley glacier, common in regions of exceptionally high relief, is fed by avalanching from snow and ice fields higher up. An especially important type of valley glacier serves to drain ice from mountain ice-caps and from ice-sheets. Fine examples occur in Iceland, Norway and Greenland.

Glaciers of piedmont type are well known from Alaska and Iceland. They form where a valley glacier advances out beyond its confining valley to expand and occupy a lowland, commonly one bordering the sea, or spreads out over the sea as an ice-tongue or ice-shelf. Piedmont glaciers are generally rather large and active. An ice thickness of several hundred metres is typical.

Floating ice-tongues and ice-shelves are at present confined to high latitudes where glaciers can reside at sea level. Ice-tongues, much smaller than ice-shelves, are normally found at the lower ends of valley and piedmont glaciers and are commonly enclosed by a prolongation of the valley walls. Ice-tongues lose mass by the calving of icebergs and by melting underneath and at the edges, though some ice-tongues calve but little, and terminate in complex lobes shaped primarily by melting. Ice-shelves are characteristic of the Antarctic and are generally thick, in excess of 500 metres. They are nourished partly by the inland ice and partly by snowfalls on the shelf itself. They lose mass by melting underneath and by calving into huge, table-like icebergs.

Ice-caps are of two sorts. Mountain ice-caps, in which the ice is typically several hundreds of metres thick, are accumulation areas, located on upland plateaux, from which outlet glaciers flow radially to lower ground. Good examples of these comparatively large ice masses are known from Iceland, Spitsbergen, the Greenland seaboard, and Patagonia. Lowland ice-caps are, on the other hand, confined to low-lying plateaux in relatively cold regions and are not particularly active, in contrast to the highly active mountain ice-caps.

They are, moreover, rather thin. Good examples of lowland ice-caps can be found on the Arctic islands west of Greenland.

The largest of all ice masses are the ice-sheets, of continental or subcontinental proportions, at present represented only in Antarctica and Greenland. The ice-sheets of Greenland and Antarctica between them account for all but about 5 per cent of the present ice. The dome-like Greenland ice-cap covers a lowland region surrounded by rugged mountains, higher on the east coast than the west, which reach an extreme altitude of 3700 metres. The ice reaches a maximum thickness of about 3300 metres, forming a high plateau, and escapes to the sea by way of numerous large outlet glaciers crossing the peripheral mountain chains. The Antarctic ice-sheet is truly vast and more complicated in nature than the Greenland ice mass. It appears to be a group of ice domes, unlike the simple dome of the Greenland ice-sheet, that surmounts a complicated terrain including several large mountain chains, whose summits here and there penetrate through the ice cover. The upper surface of the ice rises to a maximum elevation a little more than 4000 metres, whereas the ice base varies in elevation from more than 3000 metres above sea level to as low as −1200 metres in the Polar and Wilks Basins. The periphery of the Antarctic ice-sheet is marked by huge outlet glaciers, of which the Beardmore and L'Ambert are typical, and numerous ice-shelves such as the Ross and Filchner Shelves. The Filchner Shelf, lying in the Weddell Sea, has a maximum thickness of about 1400 metres and thus grounds on relatively deep parts of the sea bed.

7.4 Glacier Regime and Flow

Glaciers exist because, in the long term, ice and snow have been added faster on higher ground than ice and snow have been lost by melting and other processes on lower slopes. The term accumulation embraces all those processes whereby a glacier gains ice and snow, while the term ablation covers all those processes—melting, calving, wind erosion, evaporation—which lead to loss of mass. The balance between accumulation and ablation determines the regime of a glacier. The regime is positive if there is a net gain of mass, so that the glacier thickens and extends itself, but is negative if there is a net loss, so that thinning and retreat occur. When the mass budget is balanced, the glacier maintains its volume and position. In order to examine the regime in detail, however, we must consider the gross as well as the net rates of accumulation and ablation. It is clear that the net rates are unrelated to the gross rates, just as the rates of erosion and deposition on a stream bed are unrelated to the sediment transport rate. In general, a glacier is active and flows relatively swiftly where it experiences large gross rates of accumulation and ablation relative to its surface area. But the glacier is relatively inactive, and its rate of flow small, where these gross rates are small compared with surface area.

That glaciers move is well known, though it is impossible at present to specify the motion accurately on account of the complex physical behaviour of polycrystalline ice containing rock debris, and the spatially varying temperature regimes of natural ice masses. Some indication of patterns of movement in glaciers can be obtained from the idealized streamlines and annual stratification surfaces sketched in Fig 7.3, as well as from the crevasse patterns, of which only the more common are shown. The continuous if distorted nature of the stratification surfaces indicates a simple style of flow akin to the laminar flow of a liquid or gas.

Field studies show that glaciers flow at rates seldom less than 1×10^{-6} cm/s and rarely greater than 2×10^{-3} cm/s. The flow velocity decreases downwards from the upper surface of the glacier, though it is not necessarily zero at the bed, depending on whether or not melt-water lubricant occurs at the sole. In the case of a valley glacier, the flow velocity is greatest along the median line and least at the valley walls. Whether or not the velocity is zero at the walls depends again on the degree of lubrication of the sole. The pattern of velocity distribution in a glacier is therefore broadly similar to that in a flowing river, though the similarity does not always extend to the condition of no slip at the flow boundary.

We now know ice to be a visco-elastic substance that will not flow until a certain minimum shear stress, the elastic limit for shear displacements, is applied to it. We also know that ice as regards its viscous behaviour is non-Newtonian, i.e. the relationship between an applied shearing stress and the resulting strain is non-linear and depends on absolute stress, temperature and time. Generalizing, we may write for the flow of ice

$$\frac{du}{dy} = f(\tau - \tau_{(\mathrm{crit})}), \tag{7.1}$$

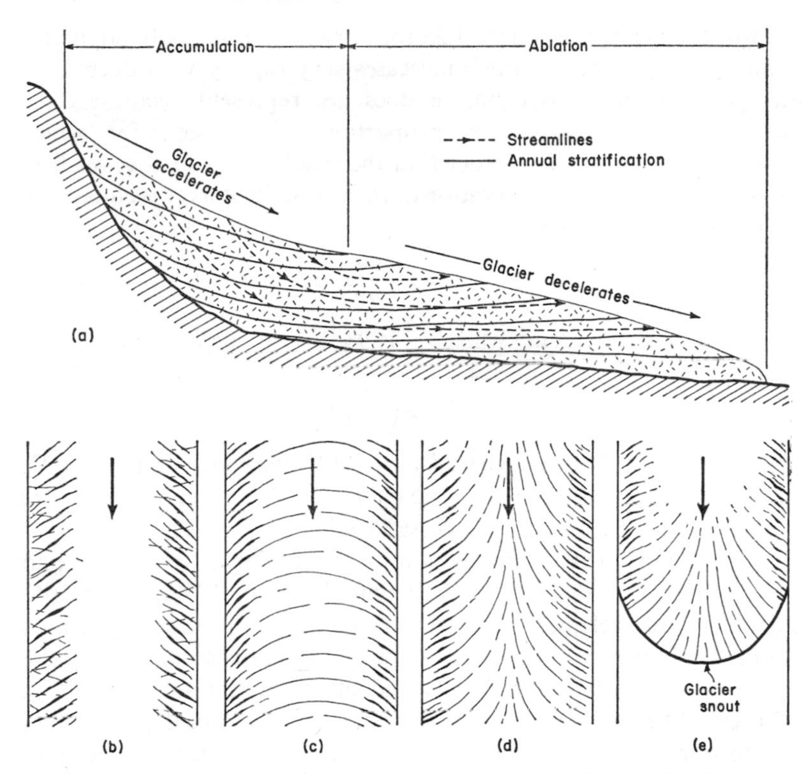

Fig 7.3 Features of glacier motion. (a) Schematic longitudinal profile through a glacier showing annual stratification surfaces and streamlines (not to be confused with particle paths). (b)-(e) Common crevasse patterns (adapted from R. P. Sharp).

in which du/dy is the velocity gradient (or strain rate), τ is the shear stress, $\tau_{(crit)}$ is the elastic limit for shear displacements, and $f(\tau - \tau_{(crit)})$ is a non-linear function. The quantity $\tau_{(crit)}$ may ordinarily be neglected in the context of glacier flow, though it is clearly important when ice is stressed rapidly over short periods, as when it is given a hammer blow, or obliged quickly to turn a sharp bend.

Controlled laboratory experiments, at strain rates on average rather larger than encountered in nature, show that the flow of ice can be represented by the equation

$$\tau = k_1 \left(\frac{du}{dy}\right)^n, \tag{7.2}$$

in which n is approximately 0·33 for $1\times10^6 < \tau < 1\times10^7$ dyn/cm^2 and k_1 is a parameter which increases very rapidly with decreasing temperature. But this equation does not represent accurately the behaviour of a significant proportion of glaciers, for which $\tau < 1\times10^6$ dyn/cm^2. In order that the relationship shall cover field as well as laboratory observations, we must write it as two equations, which are

$$\tau = k_2 \frac{du}{dy}, \tag{7.3a}$$

for $\tau < 1\times10^6$ dyn/cm^2 where k_2 is 9×10^{14} dyn s/cm^2, and

$$\tau = k_3 \left(\frac{du}{dy}\right)^n, \tag{7.3b}$$

for $\tau > 1\times10^6$ dyn/cm^2 where k_3 is 1×10^8 dyn s/cm^2 and $n \approx 0{\cdot}22$. The values of the constants and the exponent given here are appropriate for temperate glaciers. Needless to say, the 'viscosity' of ice calculated on the same basis as a Newtonian fluid is large and varies over a broad range, between about 1×10^8 and 1×10^{15} dyn s/cm^2 for tolerably large strain rates. This immediately points to the laminar nature of glacier flow, a fact that is confirmed by the patterns assumed by dirt bands and other marker surfaces within glaciers. Although these bands are commonly involved in sweeping contortions of a sometimes majestic scale and form, particularly in regions of ablation, there is nowhere any indication of a turbulent style of flow. For if we assume a glacier 500 metres thick flowing at a mean velocity of 2×10^{-3} cm/s, and assign to the ice a density of 0·9 g/cm^3 and a viscosity of 1×10^{12} dyn s/cm^2 we obtain for the flow a Reynolds number of 9×10^{-11}. This result would put the glacier flow firmly in the realm of creeping motions, when inertial forces can be entirely neglected.

In order to obtain the average shear stress exerted by a glacier on its bed, we may treat the flow exactly as we dealt with rivers in Ch. 4. We can write that

$$\tau_0 = \rho g R \sin\beta, \tag{7.4}$$

in which τ_0 is the boundary shear stress, ρ is the density of ice, R is the hydraulic radius of the glacier, and β is the declination of the glacier surface assumed parallel to the bed. Note that $R \to d$ as the flow width w grows larger compared to the flow depth d. As the bed slope is no longer small, we must now use $\sin\beta$ rather than the slope S. Since, for natural glaciers, R is of the order of tens or hundreds of metres, and β is of the order of several degrees or a few tens of degrees, it follows that glaciers exert very large forces on their beds. Calculations based on surveyed profiles and cross-sections of glaciers show that τ_0 is seldom smaller than $0{\cdot}1 \times 10^6$ dyn/cm^2 and rarely greater than $1{\cdot}6 \times 10^6$ dyn/cm^2. However, on account of the small flow velocities involved, the power of glaciers is relatively low and not much different from that of rivers.

By combining Eqs. (7.2) and (7.4) we obtain expressions for the distribution of shear stress and velocity within the body of a glacier. If the glacier flows in an infinitely wide channel of uniform depth d, we have

$$\tau = \rho g d \frac{y}{d} \sin \beta, \tag{7.5}$$

and

$$u = \frac{k_4 d(\rho g d \sin \beta)^m}{m+1}\left[1-\left(\frac{y}{d}\right)^{m+1}\right], \tag{7.6}$$

in which y is distance measured perpendicularly downwards from the upper surface of the glacier, k_4 is a constant related to k_1 in Eq. (7.2), and m is approximately 3 (i.e. $1/n$ in Eq. 7.2). We can also use these equations to give us the shear stress parallel to the wall and the velocity relative to the wall for glacier flow in an infinitely deep channel of uniform width w. We merely substitute $2z/w$ for y/d in the equations, where z is distance measured horizontally outward from the flow centre-line transversely to the direction of flow, and so obtain

$$\tau = \frac{\rho g w \sin \beta}{2} \cdot \frac{2z}{w}, \tag{7.7}$$

and

$$u = \frac{k_4 w(\frac{1}{2}\rho g w \sin \beta)^m}{2(m+1)}\left[1-\left|\frac{2z}{w}\right|^{m+1}\right]. \tag{7.8}$$

Eqs. (7.5) and (7.7) state that the shear stress varies linearly with distance from the bed or walls, while Eqs. (7.6) and (7.8) say that the velocity increases at first rapidly and then more gradually with increasing distance from the boundary.

The equations for stress and velocity in channels of other shapes are very complicated, though the general results indicated by the two pairs of expressions given above are found to be remarkably good approximations. Fig 7.4 shows plots of dimensionless values of τ and u calculated for channels of parabolic cross-section. The dimensionless quantities are defined as

$$\left.\begin{aligned} \text{dimensionless shear stress} &= T = \frac{\tau}{(\rho g d \sin \beta)} \\ \text{dimensionless velocity} &= V = \frac{u}{k_4 d(\rho g d \sin \beta)^m} \end{aligned}\right\}, \qquad (7.9)$$

where d is measured at the deepest part of the channel. Cross-sections approximating to the parabolic form are rather common amongst valley glaciers, and investigations by boring and tunneling have shown that there is a very satisfactory agreement between profiles such as those given in Fig 7.4 and the measured data. It will be noticed in the figure that the shear stress near the surface of a glacier in a parabolic channel is a maximum not at the wall itself but at a small distance inward towards the flow centre-line. One might expect to find a large number of deep, open crevasses in this zone

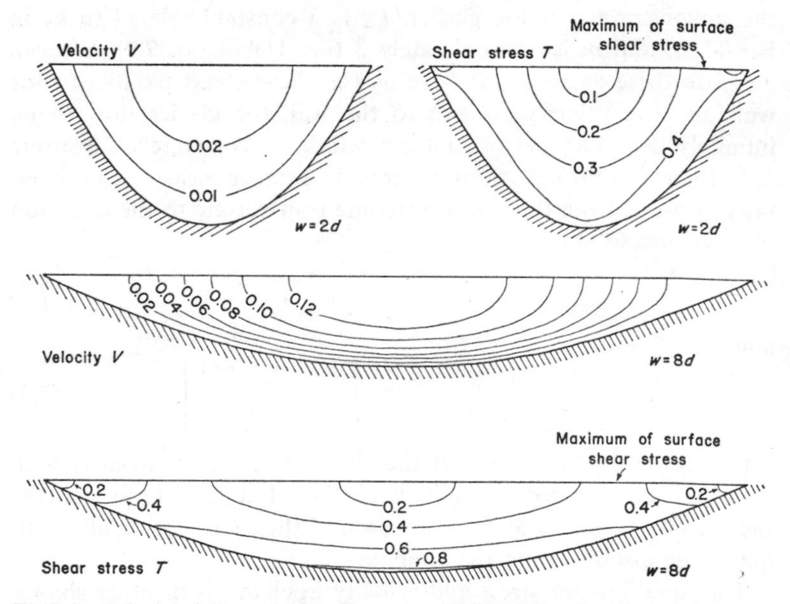

Fig 7.4 Calculated dimensionless shear stress (T) and dimensionless longitudinal velocity (V) in transverse profiles of glaciers flowing with no basal slip in parabolic channels (after J. F. Nye).

of maximum shear stress. This is precisely what is commonly observed, as sketched in Fig 7.3.

The profiles of Fig 7.4, and Eqs. (7.5) to (7.8), all depend on the supposition that $u = 0$ at the bed of the glacier. This supposition would appear to be valid for polar glaciers, whose ice seems to be held fast to the bed rock, but clearly must be relaxed in the case of temperate ones whose soles are lubricated with melt-water. With temperate glaciers there is slip between the basal ice and the bed material. The extent of the slip can be measured as a percentage of the maximum velocity of the glacier; field measurements indicate that basal slip commonly accounts for between 20 and 60 per cent of the total movement. Side slip can also be expressed as a percentage of the maximum velocity of the ice flow, and is rather more easily measured than basal slip. Measurements of side slip cover about the same range as those of basal slip. Both rates are very variable from place to place and with time.

It must be emphasized that the mechanisms involved in the basal sliding of temperate glaciers are far from being well known or understood. However, from the limited studies thus far made at glacier beds, it appears very probable, from the structure of the lowest few centimetres of the ice, that regelation is an important if not major process of basal sliding, the ice alternately melting and freezing as the local pressure conditions change. Experimentally it is possible to pass large stone blocks through ice simply by regelation.

7.5 Glacial Erosion

The erosive powers of moving ice are well attested in glaciated regions by marks indicative of abrasion displayed on rock surfaces and by various landscape forms sculptured in solid rock. The careful study of these surface markings and landscape forms has led to many valuable insights into the erosional processes of glaciers, but the knowledge thus gained is very incomplete and has still largely to be tested against, and augmented by, direct observations at the base of moving ice. Direct observation is difficult, not only because ice is a solid, but also because the very low flow speeds of glaciers call for the observation of processes over times measured in units much larger than minutes or hours.

The two processes by which a glacier directly erodes its bed are abrasion and quarrying or plucking. It is, however, impossible in the case of temperate glaciers to separate these spatially or temporally from erosional processes associated with melt-water streams that flow in tunnels between the ice and the bed rock.

Glacial erosion by abrasion involves the rubbing, polishing, scratching, grinding and scoring of the bed by mineral and rock particles serving as tools held firmly in the moving ice. From the preceding discussion of glacier motion, it is obvious that very large forces are likely to be concentrated at the points of action of such tools. In the course of abrasion, the tools themselves become worn down though, as we shall see, quarrying continually replaces them.

Striae, or striations, are the smaller surface markings that result from glacial abrasion. Striae are rectilinear, less commonly curvilinear, markings engraved in large numbers on the glaciated rock surfaces. Striae vary from a few centimetres to a few metres in length, from a fraction of a millimetre to a few centimetres in width, and from a fraction of a millimetre to several millimetres in depth. The smallest glacial striae, preserved in fine grained rocks, cannot be observed by the unaided eye. Some striae begin and end gradually, slowly deepening or shallowing along their length, while others terminate or begin abruptly, at steeply inclined fracture surfaces. One generally cannot tell from striae the direction of ice motion, only the path of flow, though when combined with data from other ice-shaped features they form one of the most useful sources of evidence on glacier flow. Since striae arise at the base of the ice, they denote the skin-friction lines of glacier motion which, like skin-friction lines for liquid and gaseous flows, are very sensitive as regards direction to the slightest irregularity on the flow boundary. Thus glacial striae are observed to diverge at cliffs that face into the direction of ice flows, and to diverge and then rejoin around knolls on the bed rock surface.

When rock or mineral tools are not held firmly in the ice, but are partly rolled along between it and the bed rock, they give rise in the surface layers of the bed rock to a variety of features which include crescentic fractures, crescentic gouges and chatter marks. These structures, like striae, depend on the local application to the bed of forces of considerable magnitude, but not in quite the same manner as when striations are being produced. Crescentic fractures are steeply inclined cracks or fracture surfaces, the outcrops of which are crescents concave in the direction of ice motion. Ordinarily the fractures are found nested in rows parallel to glacier flow. Crescentic fractures do not represent a loss of material from the bed, only its mechanical weakening. Crescentic gouges, on the other hand, do imply a loss of bed rock to the ice. They are shallow, nested, sickle-shaped hollows a few decimetres or metres wide across the wings, whose convexity either opposes or points in the direction of ice movement. In whichever direction the wings face, however, the steepest side of the gouge always opposes the direction of ice flow. Chatter marks suggest tools that moved in a stick-slip manner.

The materials which become incorporated into ice as the result of glacial abrasion are chiefly of clay, silt and sand sizes, the finer of these elements being described as 'rock flour'. Very probably abrasion provides glaciers with comparatively few if any particles of gravel size.

Glacial abrasion is most intense where the ice flow is compressed above some obstacle shaped on the bed rock surface, for example a *roche moutonée* or a valley rock bar (Fig 7.5). For in such regions the flow speed of the ice is locally increased together with the magnitude of the boundary shear stress and the force applied to the interface by a tool held in the ice. In the same regions, provided the glacier is temperate, some melting is to be expected as a consequence of the locally increased pressure. But to the lee of the obstacle, the pressure is released and abrasion must cease to be a significant process of erosion. It is, however, in the shelter of *roches moutonées* and rock bars that quarrying or plucking, the second process of glacial erosion, is found to occur (Fig 7.5). It seems very

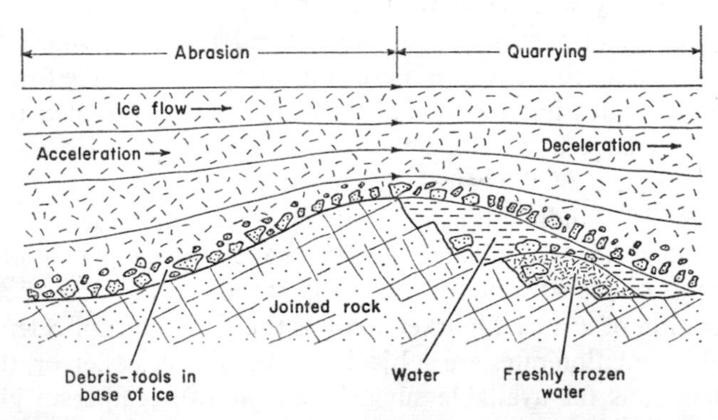

Fig 7.5 Abrasion and quarrying at an irregularit y on the rock-bed of a glacier.

likely that glacial quarrying in the lee of obstacles results from the periodic refreezing there of water formed by the melting of ice in zones where the ice flow is locally compressed. We do know from direct observations that sub-glacial cavities can exist downstream from obstacles on the bed rock surface, and that these cavities can from time to time hold water that may eventually freeze.

The quarrying probably depends specifically on the repeated freezing of melt-water in the natural joints and fracture surfaces of the rock, with the result that the joints are opened wider and wider until the joint and fracture blocks can be said to be a part of the ice flow rather than an element of the bed rock. The movement of a joint block could also be assisted by the larger tools held in the ice, though if the block is to become incorporated into the glacier, ice must be deposited on it or caused to flow around it. Quarrying provides glaciers with the coarser elements in their loads. These elements are chiefly of pebble, cobble and boulder sizes, as determined by the characteristic spacing of joints and fractures in the parent rock, but they quite commonly include huge masses or 'rafts' of rock many hundreds or thousands of cubic metres in volume.

Corrasion is the main process of erosion associated with subglacial melt-water streams, and a variety of forms sculptured in the solid rock can be attributed to it. They include large grooves cut parallel to the direction of ice motion in rock surfaces of any attitude, curved and winding channels, intricately fashioned pot-holes and bowls, and crescentic hollows whose steep sides point in the direction of ice flow. The latter, called *Sichelwannen*, are remarkably like certain flute and ripple marks fashioned by water on less resistant beds, and may well result from a differential corrasion associated with regions of separated flow in the subglacial channels. Some workers believe that cavitation erosion is important in the formation of the structures just mentioned, but this may be doubted because of the very large flow velocities that are called for.

The problem of the rate of glacial denudation has been tackled from several standpoints: the computation of the volume of direct glacial deposits, the change of landscape during glaciation, the sediment transport rate of outwash streams, and the speed of obliteration of artificial markers. All of these methods of calculating glacial denudation rates are subject to many sources of error, though the data thus far available suggest that active, temperate glaciers are potent agents of denudation. Unfortunately, data are few for polar glaciers, but it seems that glaciers of this thermal type are relatively inactive. Computations for various temperate valley glaciers and small ice-caps yield denudation rates between 1×10^3 and 1×10^4 cubic metres of bed rock per year per square kilometre of glacier bed. These rates are one or two orders of magnitude larger than are typical of river systems. However, since the power of glaciers is of the same order as that of rivers, it would appear that glaciers are the more efficient of the two denudations agents.

7.6 Glacial Deposits

Before considering glacial deposits, we should examine briefly how glaciers transport their loads. One point is here fundamental. Because glacier flow is laminar, the motion is unaccompanied by any intrinsic process whereby a particle eroded by the glacier can experience during transport a significant transverse movement away from the streamline on which lay the ice responsible for its entrainment. That is, in glacier flow there is no process corresponding to transport by convection and turbulent diffusion as found in rivers. The extent of diffusion of solid particles in ice under shear is comparable to the results of molecular diffusion in liquid flows. Over transport distances typical of valley glaciers and small ice caps, solid particles would diffuse laterally a distance of the same order as their diameter. For the same reason one would expect the glacier load to show little if any indication of sorting by particle size. There are, however, processes whereby debris can reach the heart of the moving ice, but even so, the bulk of the glacier load is restricted to ice at and close to the surface over which movement is occurring. Debris lying on the top of the glacier is often carried down deep into the ice through being washed by melt-water streams into crevasses and tunnels. Debris from the base of the ice is sometimes carried to higher positions along low-angle thrusts which are particularly common at the snouts of glaciers.

The position of the debris load relative to the moving ice is shown schematically in Fig 7.6. The bottom load consists of debris torn from the bed rock in the course of glacial abrasion or quarrying, and is confined to the lowest layers of the unfused glacial streams. Where a small and a large glacier fuse, the bottom load of the smaller may come to occupy a relatively high position within the combined ice current. The surface load, gradationally related to the bottom load, consists of material that is transported on the surface of the glacier, at the sides and, very commonly, in one or more medial streams. The surface load is polygenetic. A part represents abrasion and quarrying, but a substantial proportion consists of debris from the valley sides that has slid, tumbled or avalanched on to the glacier or been washed there by summer streams. The interior load consists of sediment that has fallen into crevasses, or been washed into tunnels, or otherwise incorporated into the central parts of the glacier.

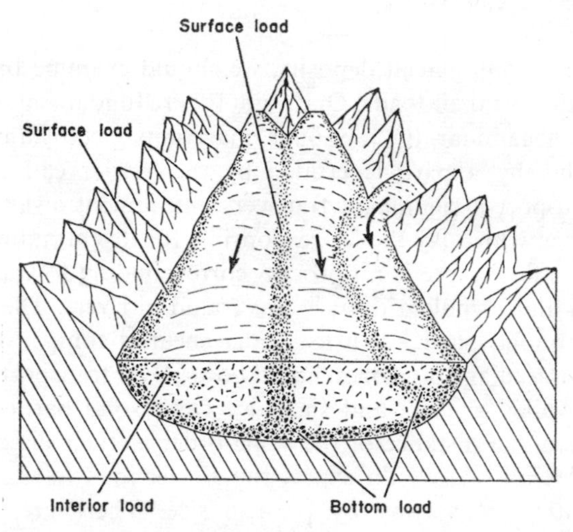

Fig 7.6 Schematic distribution of sediment loads in a glacier.

Glacial deposits, or drifts, are of two sorts: unstratified drift and stratified drift. The unstratified drifts are known generally as till, though sometimes the rather inappropriate term 'boulder clay' is given them. Tills are the only glacial deposits to be laid down directly by ice, either while in a state of motion or one of stagnation and wasting. The stratified glacial deposits are the results of stream and other processes whose action is very closely associated in space and time with the processes of the ice itself. The sites of these processes are upon, within, or immediately below the ice; the duration of their action extends from the stage of active ice movements to the end-phase of wasting. We consider only ice-contact deposits to be stratified glacial drift.

In their distal reaches glaciers lay down from the bottom load lodgement tills which take the form of irregularly thick sheets (Fig 7.7). Till sheets of this origin cover vase areas in such low-lying glaciated regions as, for example, the Interior Plains of Canada and the lowlands of Fenno-Scandia, where once ice-sheets or piedmont glaciers lay. A thickness of several tens of metres is typical of till sheets.

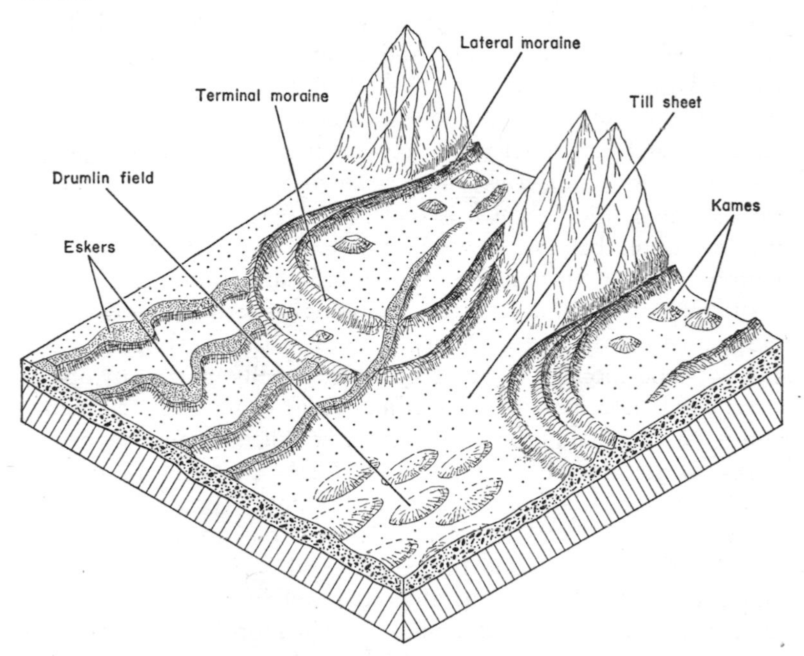

Fig 7.7 Schematic morphological and stratigraphical expression of the main types of glacial deposits.

The top surface of a till sheet is seldom even, and is commonly ornamented with regularly spaced and arranged mounds and hollows to which the term moraine may be applied. These features are of uncertain origin but they do suggest an unstable mode of behaviour for ice or debris flowing near the base of a glacier. A very common pattern consists of rounded ridges a few metres in height, between 100 and 300 metres in spacing, and up to 2·5 kilometres long, that lie transversely to the direction of ice motion. In some cases, faint longitudinal ridges are associated with the transverse pattern. Elsewhere on a till sheet, much bolder but still regularly spaced transverse elements or ribs are seen, sometimes with fainter longitudinal structures. The ribs are till ridges between 10 and 30 metres high, up to 5 kilometres along the crest, and with crests 100 to 300 metres apart. Instead of a transverse pattern a till sheet may display morainic features elongated parallel to the direction of glacier flow. There is as regards the longitudinal features a gradation between glacial flutings and drumlins by way of drumlinoid ridges. Glacial flutings are evenly spaced, parallel, curvilinear ridges and hollows with a relief of a few metres and a length of many kilometres. Drumlins are streamlined mounds of elliptical to lemniscate plan that occur in groups and comprise one of the most familiar of morainic land forms. They are between 10 and 50 metres high and between 1 and 5 kilometres long. The length is between 2 and 5 times the maximum width, which is commonly measured a little nearer the upcurrent than the downcurrent end. Drumlinoid ridges are morphologically intermediate between drumlins and glacial flutings.

In contrast to till sheets, the formation of terminal and related moraines is associated with processes of ablation (Fig 7.7). The successive positions of a glacier snout are usually marked by one or more curvilinear moraines, sometimes rising more than 150 metres above the surrounding terrain, that may be called terminal moraines. Terminal moraines have a gradational relationship with the less boldly developed lateral and medial moraines. Terminal moraines include material from the bottom load of the glacier and also a considerable amount from the surface and interior loads. Lateral and medial moraines are generally rather rich in interior and surface load debris. Terminal and similar moraines ordinarily show very irregular crests which often suggest a row of piles of debris rather than a smoothly continuous ridge. It is not uncommon to find that the surface of a till sheet, particularly near to terminal moraines, is dotted with irregular hummocks of ablation moraine.

Following the occurrence of melting underneath an ice tongue or shelf and from bergs and flows, a glacier can, under appropriate conditions, deposit some or all of its load at sea (Fig 7.8). Where deposition occurs appears to depend on the thermal regime of the glacier. Deposition from a polar glacier begins close to the seaward margin of the buoyant ice, and extends out beneath the iceberg zone. The temperate glacier, on the other hand, deposits a till sheet which dips down below sea level to the point where the ice first floats, while from the buoyant ice there is a continual downward

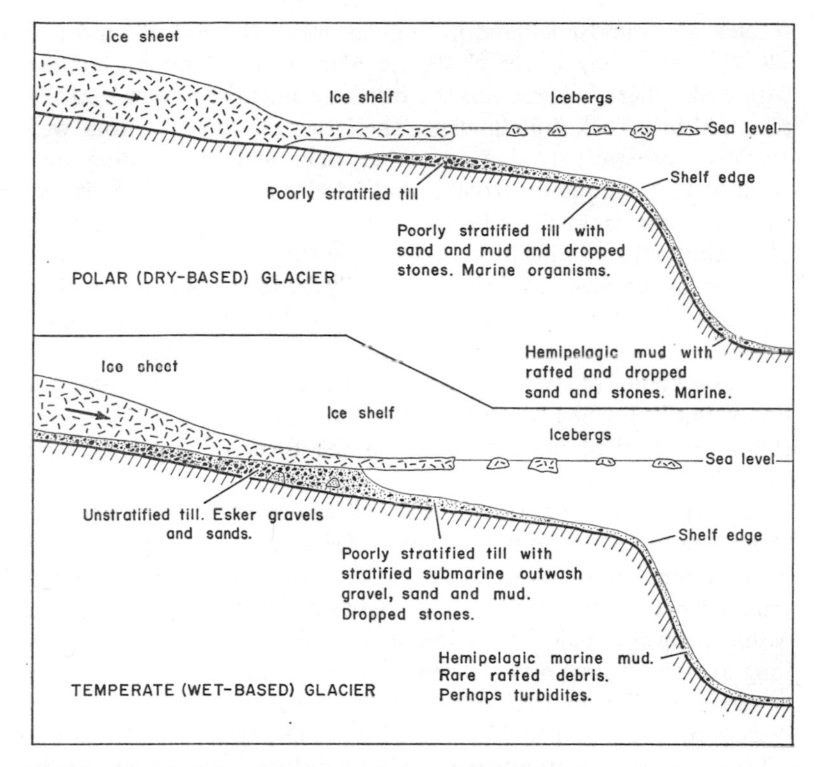

Fig 7.8 Model of glacial deposition from partly floating glacier ice, as controlled by thermal regime (after H. G. Reading and R. Walker).

rain of debris. The scale of these systems should be noticed. The region of buoyant ice is commonly several tens or hundreds of kilometres wide; icebergs and floes stray as much as 3000 kilometres in a direction normal to the ice cliffs from which they broke away, and thus reach well beyond the edges of the continental shelves.

As sediments, tills are as varied as their parent rocks. A complete absence of sorting, with a range of particle sizes from a few microns to several metres, is perhaps characteristic of the great tills as found in till sheets derived from mixed hard and soft bed rocks, though sediments of other origins can be almost as ill sorted, for example, mud flow deposits. In tills, however, there is rarely a significant amount of clay, except in those derived from a mainly shaly terrain. The bulk of material in till consists of chemically fresh 'rock flour', in which labile minerals, such as feldspar and even ferromagnesian species, are found unaltered in angular grains the size of sand and silt. Generally, the larger clasts are numerous but not in contact with each other; in some tills the outsize elements are very sparsely scattered in a rock flour matrix. The larger clasts are seldom well rounded, generally having no more than smoothed edges and corners, but commonly striated and locally faceted and polished. Most tills are compact and unstratified though a few show a rude, often contorted stratification. Low angle thrust surfaces give some tills the appearance of being internally bedded. Numerous tills, especially those formed by ablation, enclose small bodies of stratified sediment of ice-contact origin. Because of current action, rather extensive bodies of stratified sediment are to be expected in association with tills formed beneath floating ice. The drop of rafted stones should be evident in these cases in the form of depressed and ruptured laminations.

Neither the matrix grains nor the outsize elements of till ordinarily show a random orientation in space. Usually the long axes of the greatest number of particles lie close to one direction, while the remainder take up positions at right angles to the orientation assumed by the majority. A low angle imbrication of the particle long axes can sometimes also be measured. Thus the fabric of tills is rather similar to that observed from water-laid sediments, as discussed in Ch. 2. The fabric of tills is interpreted, mainly on the evidence of other features due to movement, to mean that the bulk of the till debris aligns itself during transport and deposition parallel to the direction of ice motion. This is probably true in the case of lodgement tills, but must be accepted with care for tills formed by ablation, where the fabric may depend more on ice melting and downslope movement, and not uncommonly on ice thrusting. Till fabrics are now known to be much more variable statistically than was at first thought, and many early conclusions based on them must be treated cautiously. The fabric probably depends on the forces exerted by the ice, the ice content of the till, and the shape, size and concentration of the particles of different sizes. The origin of till fabric is far from being properly understood.

The most important of ice-contact deposits are those found as kames, kame terraces and eskers (Fig 7.7). Most of these bodies of sediment were formed by stream action in or upon the ice, and were subsequently let down on to the ground when the ice melted away. Though consisting chiefly of stratified sediment, the characteristic feature of kame, kame terrace, and esker deposits is evidence for collapse, in the form of relatively large scale contortions of the bedding and multiple faults, especially at the margins of the bodies.

Eskers are sinuous bodies of sediment laid down by streams flowing in tunnels within or at the base of the ice. They are typically between 5 and 50 metres high, between 15 and 250 metres across, and from 500 metres to 200 kilometres long. Multitudes of large and small eskers are associated with the great till sheets of Fenno-Scandia and Canada. Internally, eskers consist of rudely stratified boulder gravels with some cross-bedded or evenly laminated sands. Large, cross-bedded, delta-like bodies of gravelly sediment can be found here and there along the length of many eskers, and presumably these were formed where the ice tunnel expanded in cross-section to give a chamber and pool. The upper layers of many eskers consist, as in Sweden, of esker material reworked and sorted by wave and current action in a lake or sea. Beach terraces, associated with well sorted and rounded gravel, occur on the tops and sides of these eskers.

Kames are mound-like hills of generally fairly well sorted, but usually only poorly stratified, gravel and sand. They represent collapsed crevasse infillings or supraglacial lake deltas, and in the latter case varved clays, generally showing evidence of collapse, may be found with them. A kame terrace is formed where stream deposits accumulated at the edge of the glacier collapse. Kame terraces often show kettles, which are holes left by the melting of *remanié* ice blocks. The terrace deposits are well bedded sands and gravels.

7.7 Ancient Glacial Sediments

The claim of a glacial origin has been made for very many bouldery deposits of several different pre-Pleistocene ages scattered over the five continents. These claims are not easily tested conclusively, as ill-sorted, boulder-bearing sediments can arise by several non-glacial processes, for example mud flow or submarine sliding, and striated surfaces and clasts are not restricted to the glacial environment. However, a careful study of these ancient deposits with the Pleistocene glacial sediments in mind, suggests that many claims to a glacial sediment are false and exaggerated. The evidence for glacigenous sediment is firm at four main dates: Precambrian, Late Precambrian, Devonian and Permo-Carboniferous. In view of the correlation between glaciation and high latitude, ancient glacial deposits have for many years attracted the interest of workers concerned with the nature and extent of polar and continental wandering.

The Precambrian glaciation, occurring somewhere between 1500 and 1100 million years ago, is represented by several boulder beds in the rocks of the Canadian Shield, from which some of the greatest of the Pleistocene till sheets were later derived. Amongst these boulder beds is the Gowganda tillite, extending over an area of about 500 000 km^2 and associated locally with a striated pavement.

Glacigenous sediments of Late Precambrian age have been recorded from nearly every continent, and their wide distribution and high incidence suggests a Late Precambrian glaciation, or more probably a series of glaciations, of great extent and severity. Many of the tillite beds extend laterally over hundreds of kilometres, though rarely exceeding a few tens of metres in thickness. Some Late Precambrian tillites appear to be of marine origin, to judge from the evidence of dropped stones, whereas others lie on scored pavements and were deposited probably by terrestrial ice. In the Kimberley district of Western Australia they rest on smoothed rock surfaces showing striae, grooves, crescentic gouges and chatter marks.

From Brazil and Argentina there are scattered records, some from deep drillings, of widespread but thin tillites of Lower and Upper Devonian ages. Some of the tillites at outcrop rest on glaciated pavements.

The Permo-Carboniferous tillites of the Southern Hemisphere are perhaps the most discussed of all ancient glacial sediments. They are found in South America (Brazil), southern Africa, India and Australia. Amongst them are the well known Dwyka tillite of southern Africa and the Kuttung tillite of Australia, both of which rest on striated pavements with *roches moutonées*s and excavated valleys. The former is associated with varved clays and appears to be shaped into drumlin-like features. In central India, the Talchir boulder bed, about 15 metres thick, rests on a beautifully striated and quarried rock surface.

READINGS FOR CHAPTER 7

The following, in order of increasing depth of treatment, are valuable as reference works on glacial processes, geomorphology, and geology:

WEST, R. G. 1968. *Pleistocene Geology and Biology*. Longmans Green and Co., London, 377 pp.

EMBLETON, C. and KING, C. A. M. 1968. *Glacial and Periglacial Geomorphology*. Edward Arnold Ltd., London, 608 pp.
LLIBOUTRY, L. 1964. *Traité de Glaciologie*. Masson et Cie, Paris, 1040 pp.

A description of a glacier worthy of especially detailed study has been made by:

SHARP, R. P. 1958. 'Malaspina Glacier, Alaska.' *Bull. geol. Soc. Am.*, **69**, 617–646.

Studies of the physical properties and behaviour of ice have been made by numerous workers in different fields, and the results are somewhat scattered. The following illustrate the methodology of these studies and the major results thus far obtained:

GLEN, J. W. 1955. 'The creep of polycrystalline ice.' *Proc. R. Soc.*, A **228**, 519–538.
KAMB, B. and LACHAPELLE, E. 1964. 'Direct observation of the mechanisms of glacier sliding over bedrock.' *J. Glaciol.*, **5**, 159–172.
NYE, J. F. 1952. 'The mechanisms of glacier flow.' *J. Glaciol.*, **2**, 82–93.
NYE, J. F. 1965. 'The flow of a glacier in a channel of rectangular elliptical or parabolic cross-section.' *J. Glaciol.*, **5**, 661–690.
PATERSON, W. S. B. 1969. *The Physics of Glaciers*. Pergamon Press, Oxford, 250 pp.
SHARP, R. P. 1954. 'Glacier flow: a review.' *Bull. geol. Soc. Am.*, **65**, 821–838.
SHARP, R. P. 1960. *Glaciers*, The Condon Lectures, Oregon State System of Education, Eugene, Oregon, 78 pp.

There has been a strong tendency amongst workers to adopt a rather specialized approach to the lithological features of glacial sediments, and the reference works cited above provide by far the most satisfactory overall view. However, the following studies have useful features or are representative:

CONOLLY, J. R. and EWING, M. 1965. 'Ice-rafted detritus as a climatic indicator in Antarctic deep-sea cores.' *Science*, **150**, 1822–1824.
ELSON, J. A. 1961. 'The geology of tills.' *Proc. Fourteenth Canadian Soil Mechanics Conf.*, Nat. Research Co., Ottawa, p. 5–17.
HARRISON, P. W. 1957. 'A clay-till fabric: its character and origin.' *J. Geol.*, **65**, 275–306.
HOUGH, J. L. 1950. 'Pleistocene lithology of Antarctic ocean-bottom sediments.' *J. Geol.*, **58**, 254–260.
HOUGH, J. L. 1956. 'Sediment distribution in the southern oceans and Antarctica.' *J. sedim. Petrol.*, **26**, 301–306.
PREST, V. K. 1968. 'Nomenclature of moraines and ice-flow features as applied to the glacial map of Canada.' *Geol. Surv. Pap. Can. 67-57*, 32 pp.
SITLER, R. F. 1963. 'Petrology of till from northeastern Ohio and Pennsylvania.' *J. sedim. Petrol.*, **33**, 365–379.

The paper of Elson, although not readily accessible, is an especially comprehensive view of till lithology.

A discussion of how to identify ancient tills is given by:

HARLAND, W. B., HEROD, K. N. and KRINSLEY, D. H. 1966. 'The definition and identification of tills and tillites.' *Earth-Science Reviews*, **2**, 225–256.

Instructive accounts of pre-Pleistocene glacial and associated sediments are published by:

BANERJEE, I. 1966. 'Turbidites in a glacial sequence: a study of the Talchir Formation, Raniganj Coalfield, India.' *J. Geol.*, **74**, 593–606.

FRAKES, L. A. and CROWELL, J. C. 1967. 'Facies and paleogeography of Late Palaeozoic diamictite, Falkland Islands.' *Bull. geol. Soc. Am.*, **78**, 37–58.

HAMILTON, W. and KRINSLEY, D. 1967. 'Upper Paleozoic glacial deposits of South Africa and Southern Australia.' *Bull. geol. Soc. Am.*, **78**, 783–800.

PERRY, W. J. and ROBERTS, H. G. 1968. 'Late Precambrian glaciated pavements in the Kimberley region, Western Australia.' *J. geol. Soc. Aust.*, **15**, 51–56.

READING, H. G. and WALKER, R. G. 1966. 'Sedimentation of Eocambrian tillites and associated sediments in Finmark, northern Norway.' *Palaeogeog. Palaeoclimatol. Palaeoecol.*, **2**, 177–212.

Appendixes

APPENDIX I. Physical quantities and their dimensions

Quantities	Units	Dimensional formulae
Geometrical		
Length	cm	L
Particle diameter	cm	L
Water or channel depth	cm	L
Boundary layer thickness	cm	L
Area	cm^2	L^2
Volume	cm^3	L^3
Kinematical		
Time	s	T
Velocity	cm/s	LT^{-1}
Acceleration	cm/s^2	LT^{-2}
Diffusivity (general)	cm^2/s	L^2T^{-1}
Kinematic viscosity (a diffusivity)	cm^2/s	L^2T^{-1}
Rate of fluid discharge	cm^3/s	L^3T^{-1}
Dynamical		
Mass	g	M
Point force	dyn (dyne)	MLT^{-2}
Density	g/cm^3	ML^{-3}
Specific weight	g wt/cm^3	$ML^{-2}T^{-2}$
Dynamic viscosity	P (poise)	$ML^{-1}T^{-1}$
Sediment transport rate	g/cm width s	$ML^{-1}T^{-1}$
Pressure	dyn/cm^2	$ML^{-1}T^{-2}$
Shear stress	dyn/cm^2	$ML^{-1}T^{-2}$
Momentum	g cm/s	MLT^{-1}
Energy	erg	ML^2T^{-2}
Work	erg	ML^2T^{-2}
Erosion or deposition rate (of sediment)	g/cm^2 s	$ML^{-2}T^{-1}$
Power	erg/s	ML^2T^{-3}
Power as exerted by a fluid stream	erg/cm^2 s	MT^{-3}

APPENDIX II. Classification of sedimentary structures (not exhaustive)

EXOGENETIC				ENDOGENETIC	BIOGENETIC
BED FORMS		SURFACE MARKINGS	INTERNAL STRUCTURES		
COHESIONLESS BEDS	COHESIVE BEDS				
Antidunes	Flute marks	Bubble impressions	Cross-bedding	Ball and pillow	Feeding burrows
Barkhan dunes	Meandering grooves and channels	Gas pits	Cross-lamination	Convolute laminations	Living burrows
Current ripples	Mud mounds and hollows	Rain prints	Horizontal lamination or flat-bedding	Load casts	Tracks
Dome-shaped dunes	Transverse scour marks	Sand volcanoes	Scour and channel fills	Sandstone balls	Trails
Longitudinal dunes		Sun cracks		Slides	
Meander bars				Slumps	
Parting lineations					
Pyramidal dunes					
Sand ribbons					
Tidal current ridges					
Transverse dunes					
Wave-current ripples					

OTHER STRUCTURES: Shear wrinkles, tool marks.

APPENDIX III. Mathematical signs and abbreviated forms

Sign or Form	Meaning
$\bar{x}$	An average value of x
$\lvert x \rvert$	The value of x to be treated as positive
$x = y$	x equal to y
$x \approx y$	x approximately equal to y
$x > y$	x greater than y
$x \geqslant y$	x equal to or greater than y
$x < y$	x less than y
$x \leqslant y$	x equal to or less than y
$x \rightarrow y$	x tends in value to y
$x \propto y$	x is proportional to y
dy/dx	Differential coefficient of y with respect to x
$\partial y/\partial x$	Partial differential coefficient of y with respect to x
$\sin\theta$	Sine of the angle θ
$\cos\theta$	Cosine of the angle θ
$\tan\theta$	Tangent of the angle θ
$\sinh\theta$	Hyperbolic sine of the angle θ
$\cosh\theta$	Hyperbolic cosine of the angle θ
$\tanh\theta$	Hyperbolic tangent of the angle θ

INDEX